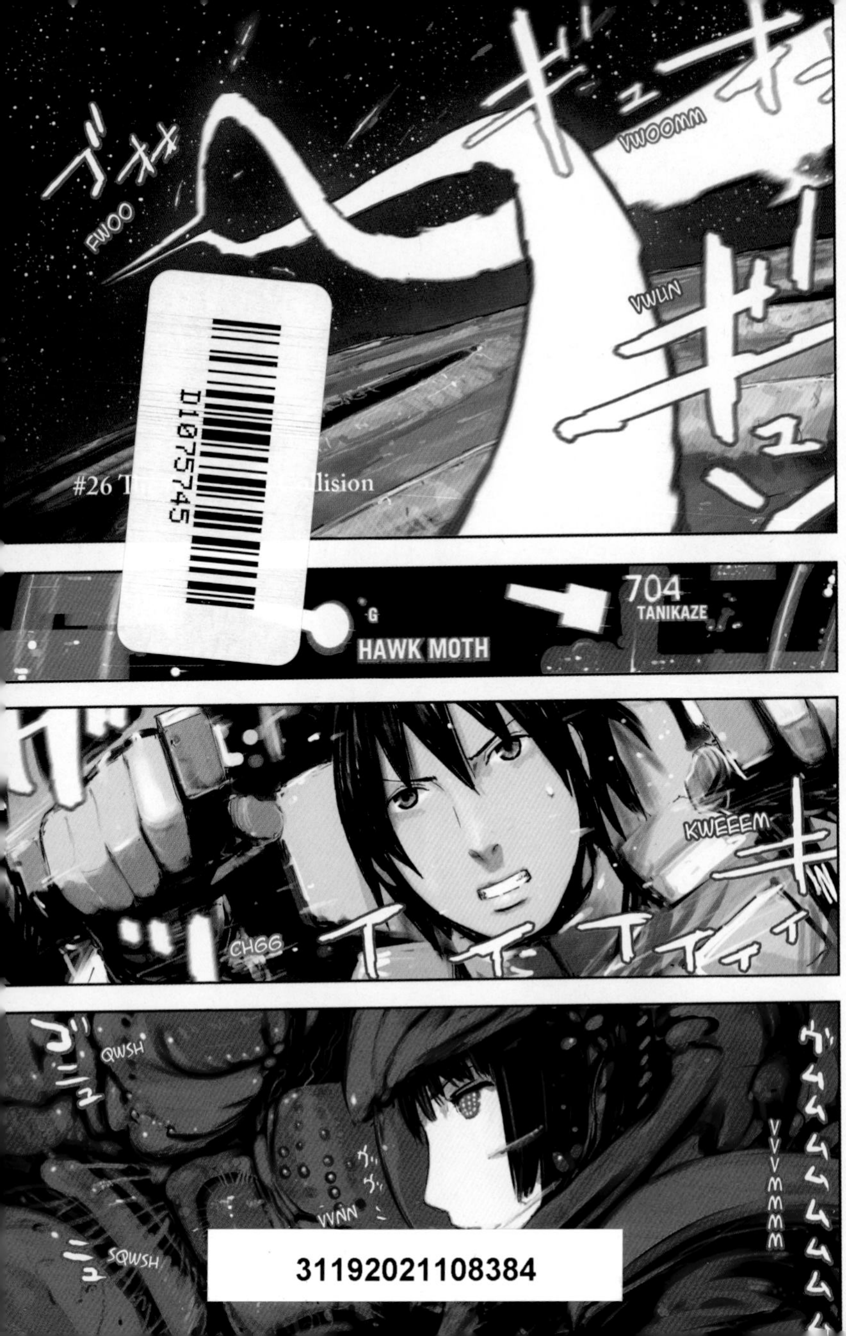
VWOOMM
VWUN
FWOO
#26 Th
llision
704
TANIKAZE
G
HAWK MOTH
KWEEEM
CHGG
VVVMMMM
VVNN
QWSH
SQWSH

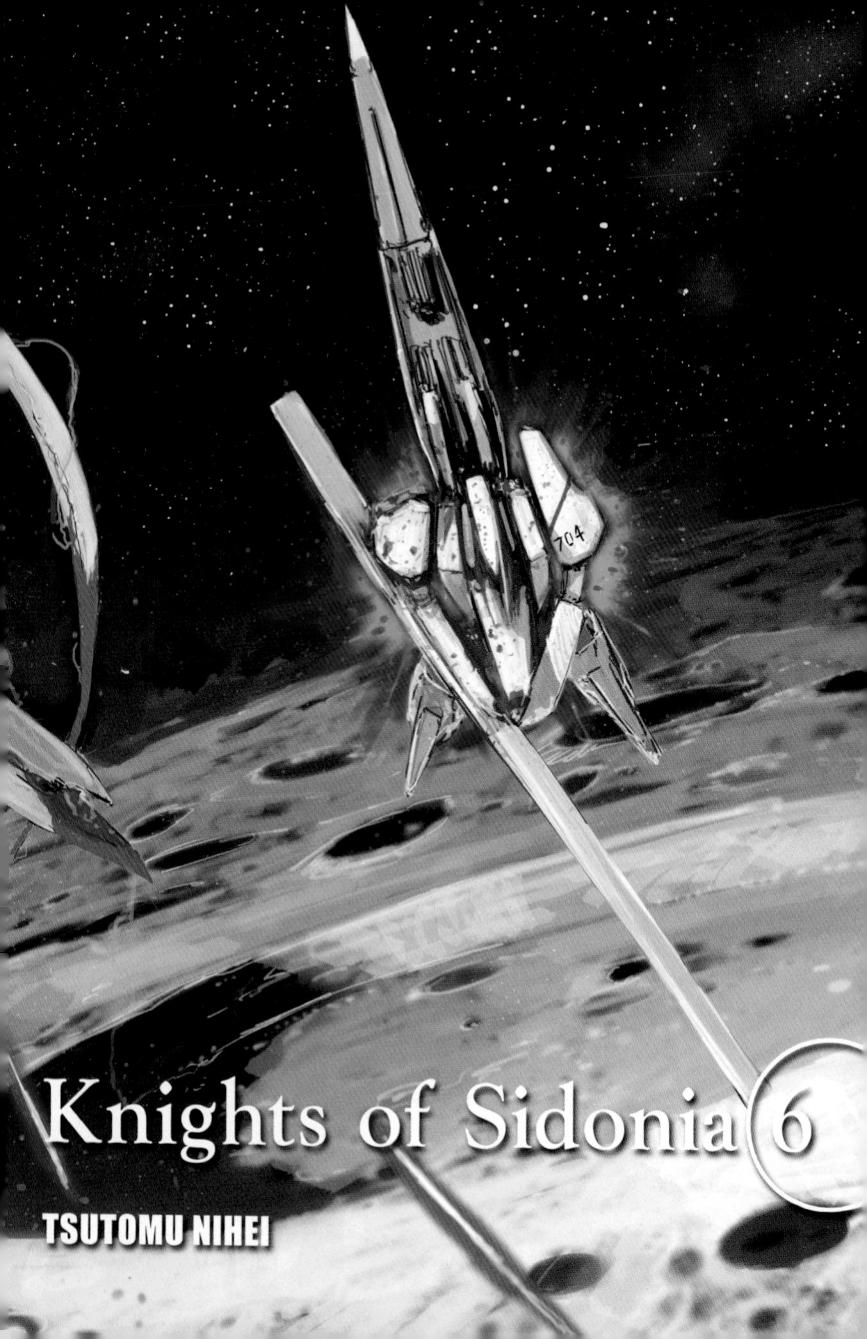
704
Knights of Sidonia 6
TSUTOMU NIHEI

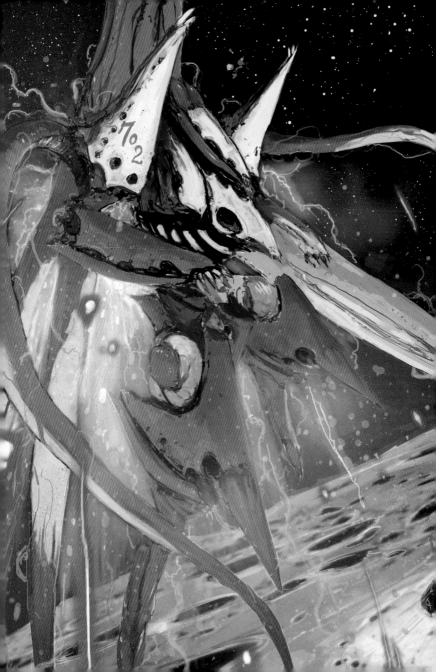

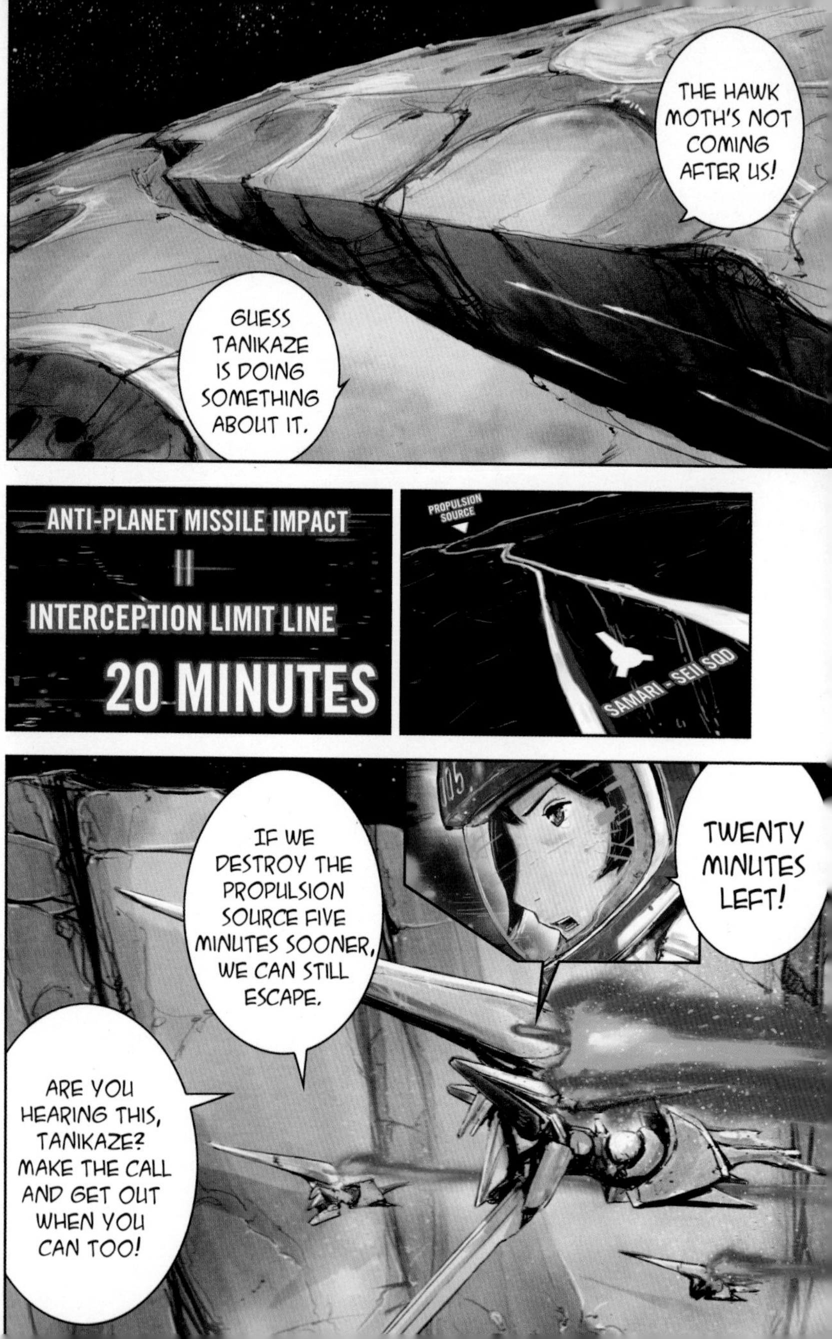
THE HAWK MOTH'S NOT COMING AFTER US!
GUESS TANIKAZE IS DOING SOMETHING ABOUT IT.
PROPULSION SOURCE
SAMARI - SEII SQD
ANTI-PLANET MISSILE IMPACT
II
INTERCEPTION LIMIT LINE
20 MINUTES
TWENTY MINUTES LEFT!
IF WE DESTROY THE PROPULSION SOURCE FIVE MINUTES SOONER, WE CAN STILL ESCAPE.
ARE YOU HEARING THIS, TANIKAZE? MAKE THE CALL AND GET OUT WHEN YOU CAN TOO!

HIGH HIGGS ACTIVITY IN THE VICINITY OF THE PROPULSION SOURCE !!

WHAT IS THAT ?!!

?!

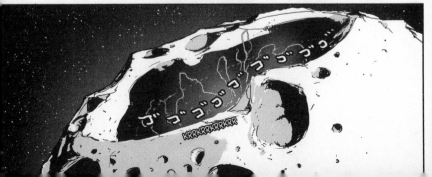

ゴゴゴゴゴゴゴ

KRRKRRKRRKRR

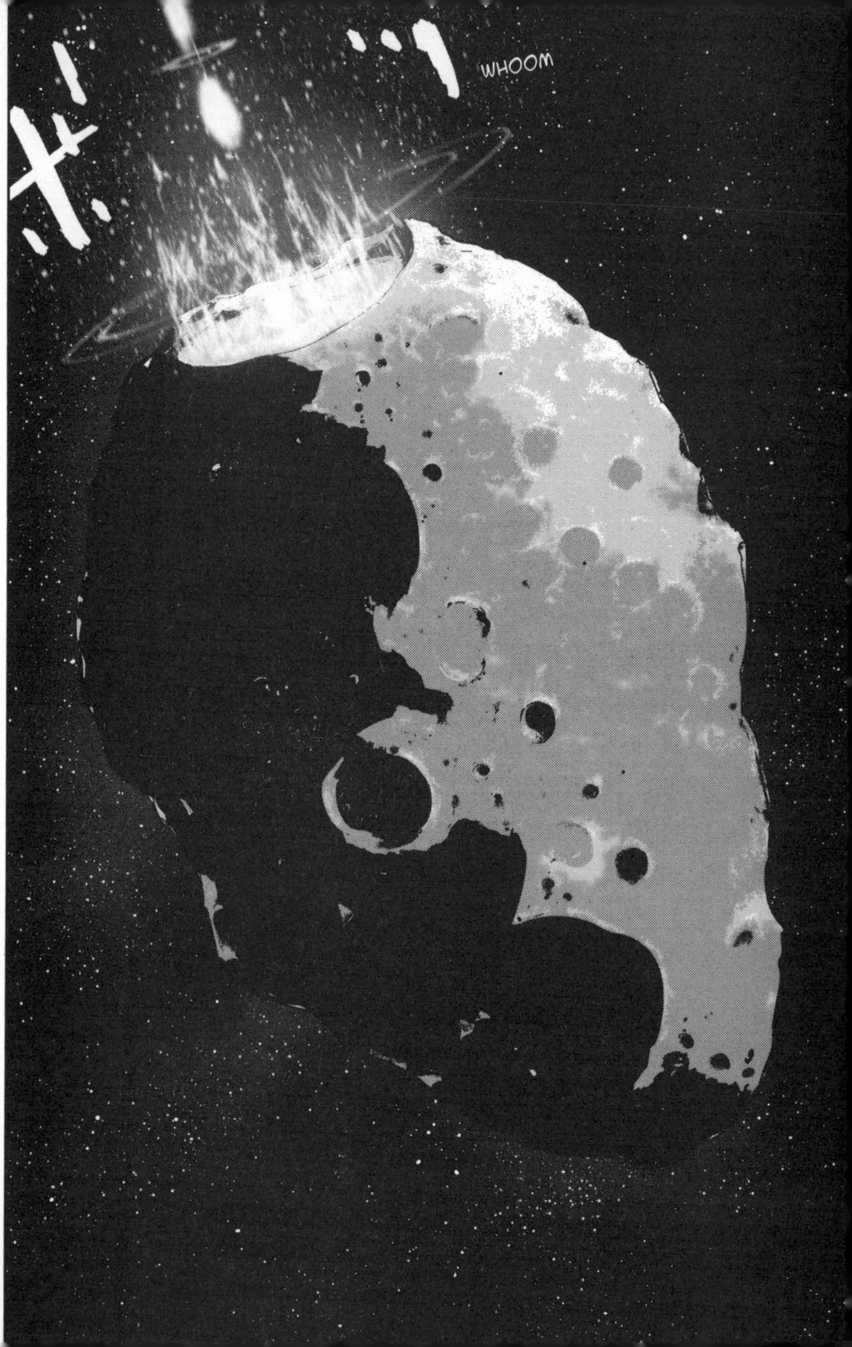
WHOOM

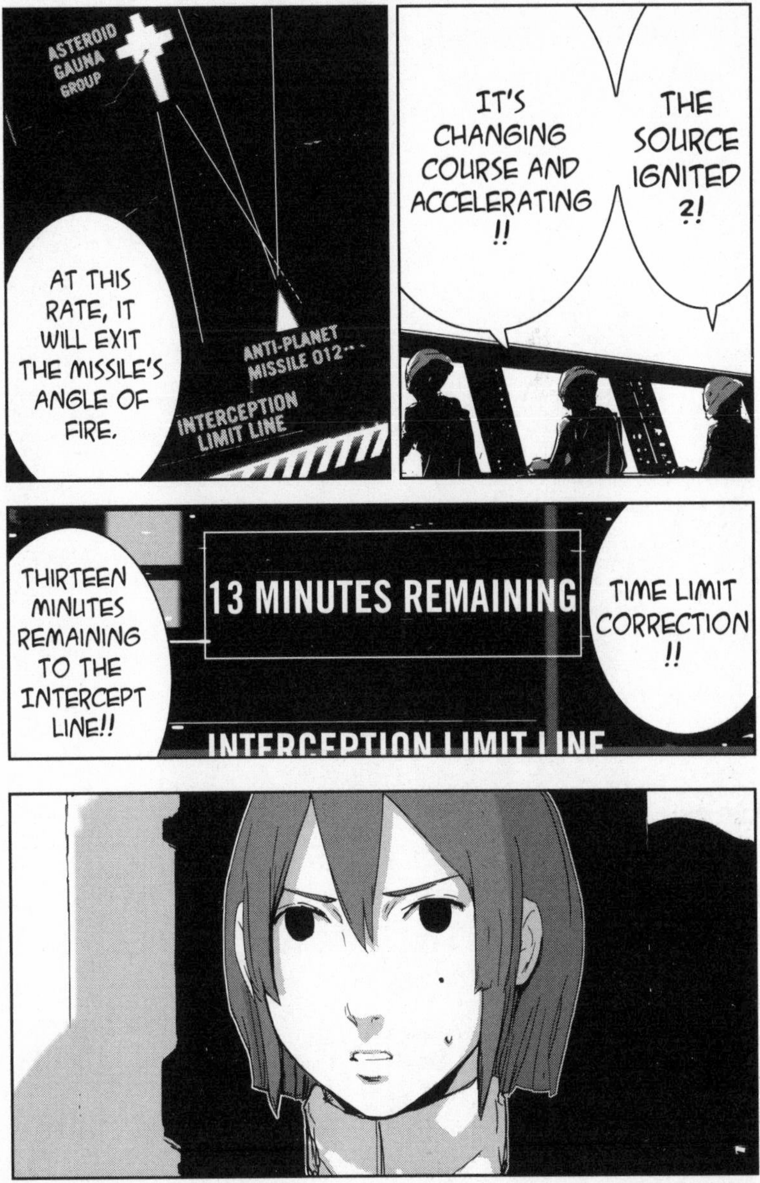
THE SOURCE IGNITED ?!
IT'S CHANGING COURSE AND ACCELERATING !!
ASTEROID GAUNA GROUP
ANTI-PLANET MISSILE 012
INTERCEPTION LIMIT LINE
AT THIS RATE, IT WILL EXIT THE MISSILE'S ANGLE OF FIRE.
TIME LIMIT CORRECTION !!
13 MINUTES REMAINING
THIRTEEN MINUTES REMAINING TO THE INTERCEPT LINE!!
INTERCEPTION LIMIT LINE

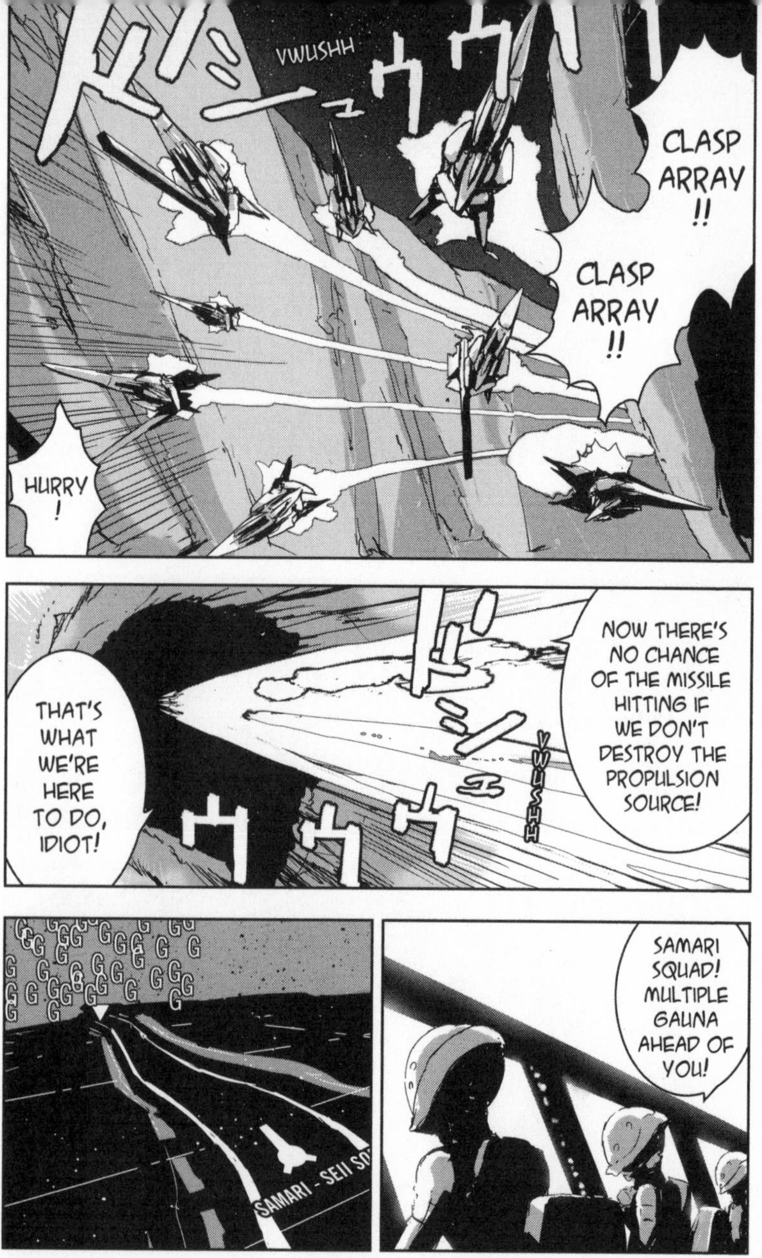
CLASP ARRAY !!
CLASP ARRAY !!
VWUSHH
HURRY !
NOW THERE'S NO CHANCE OF THE MISSILE HITTING IF WE DON'T DESTROY THE PROPULSION SOURCE!
VWUSHH
THAT'S WHAT WE'RE HERE TO DO, IDIOT!
SAMARI SQUAD! MULTIPLE GAUNA AHEAD OF YOU!
SAMARI - SEII

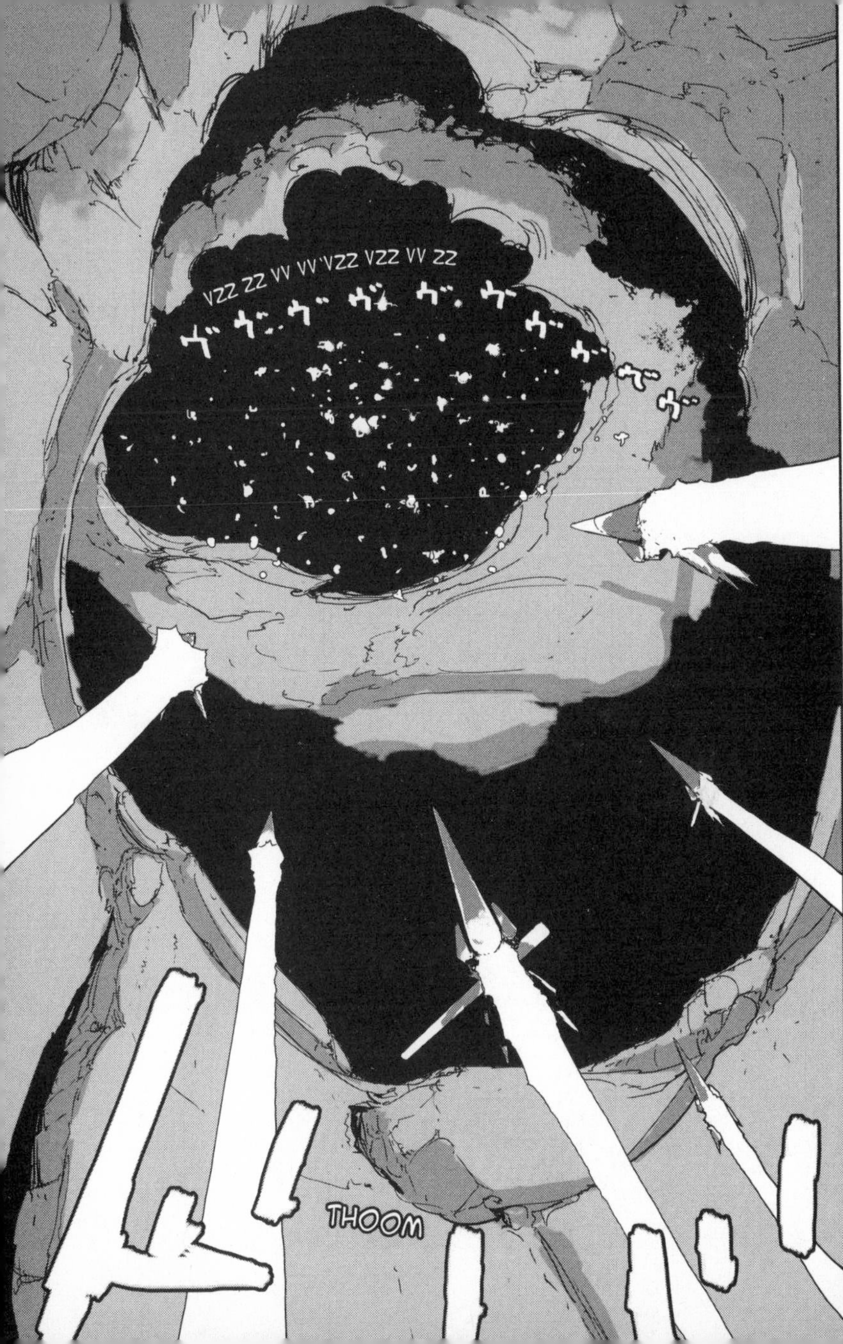
VZZ ZZ VV VV VZZ VZZ VV ZZ
THOOM

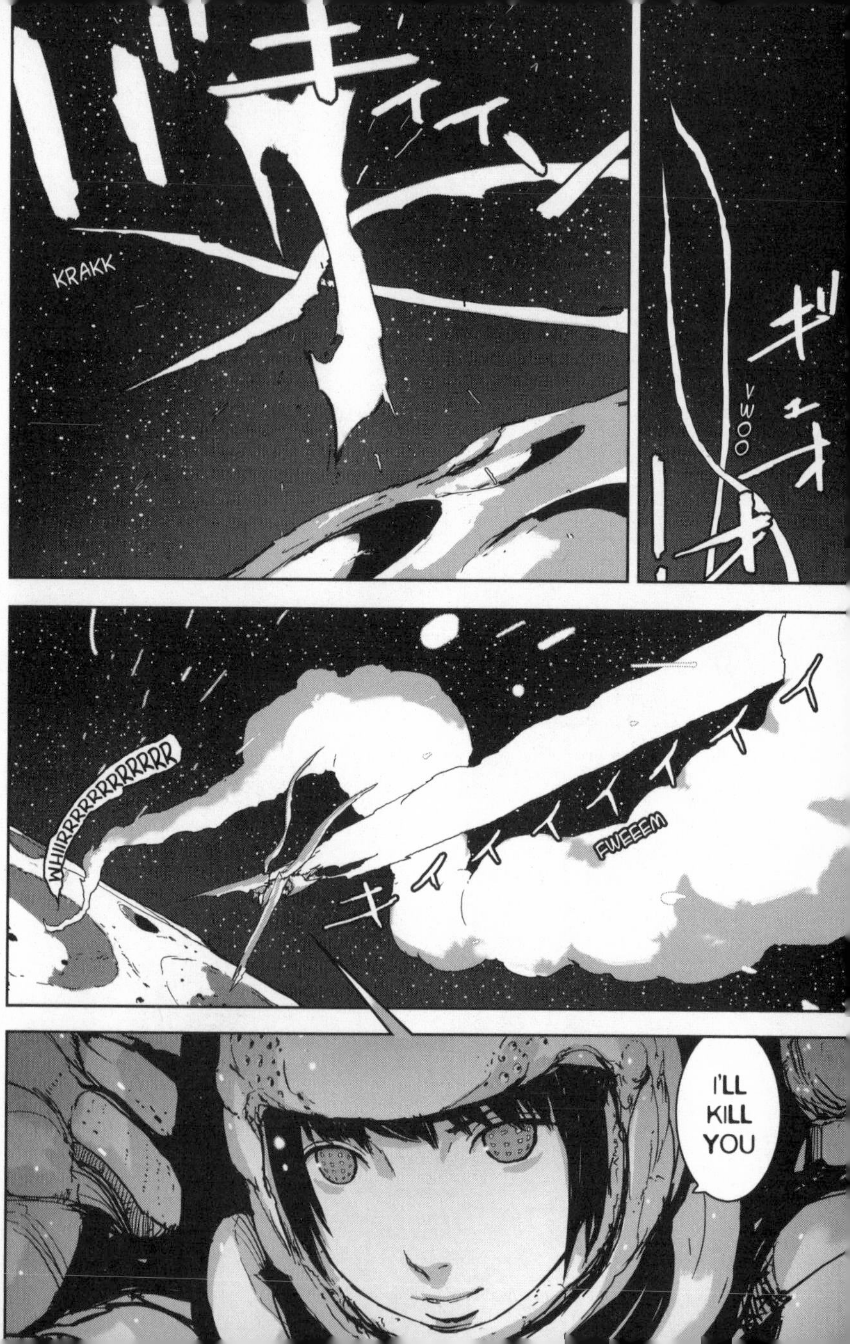
VWOO
KRAKK
FWEEEM
WHIIIRRRRRRRRRRRR
I'LL KILL YOU

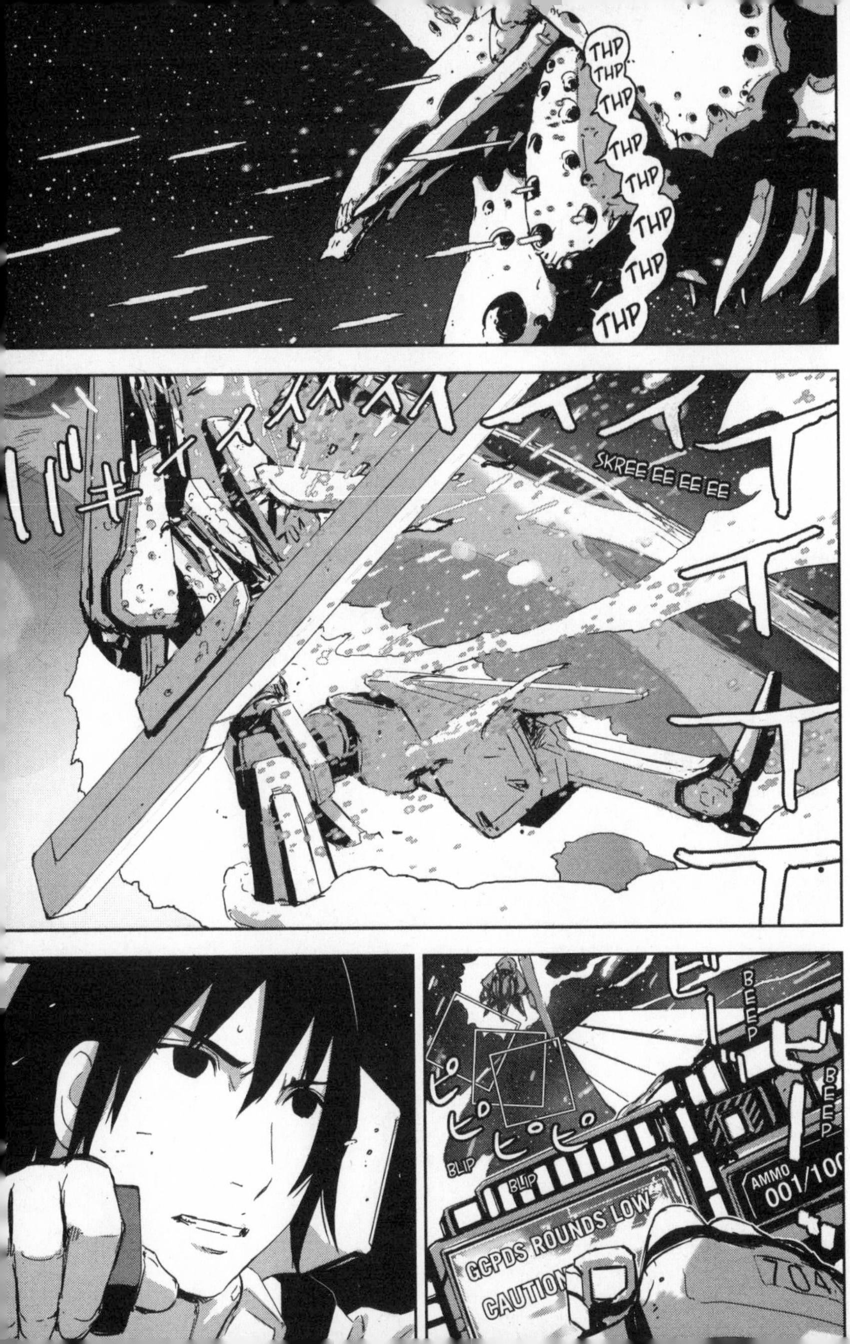
THP
THP...
THP
THP
THP
THP
THP
THP
SKREE EE EE EE
BEEP
BEEP
BLIP
BLIP
AMMO
001/100
GCPDS ROUNDS LOW
CAUTION

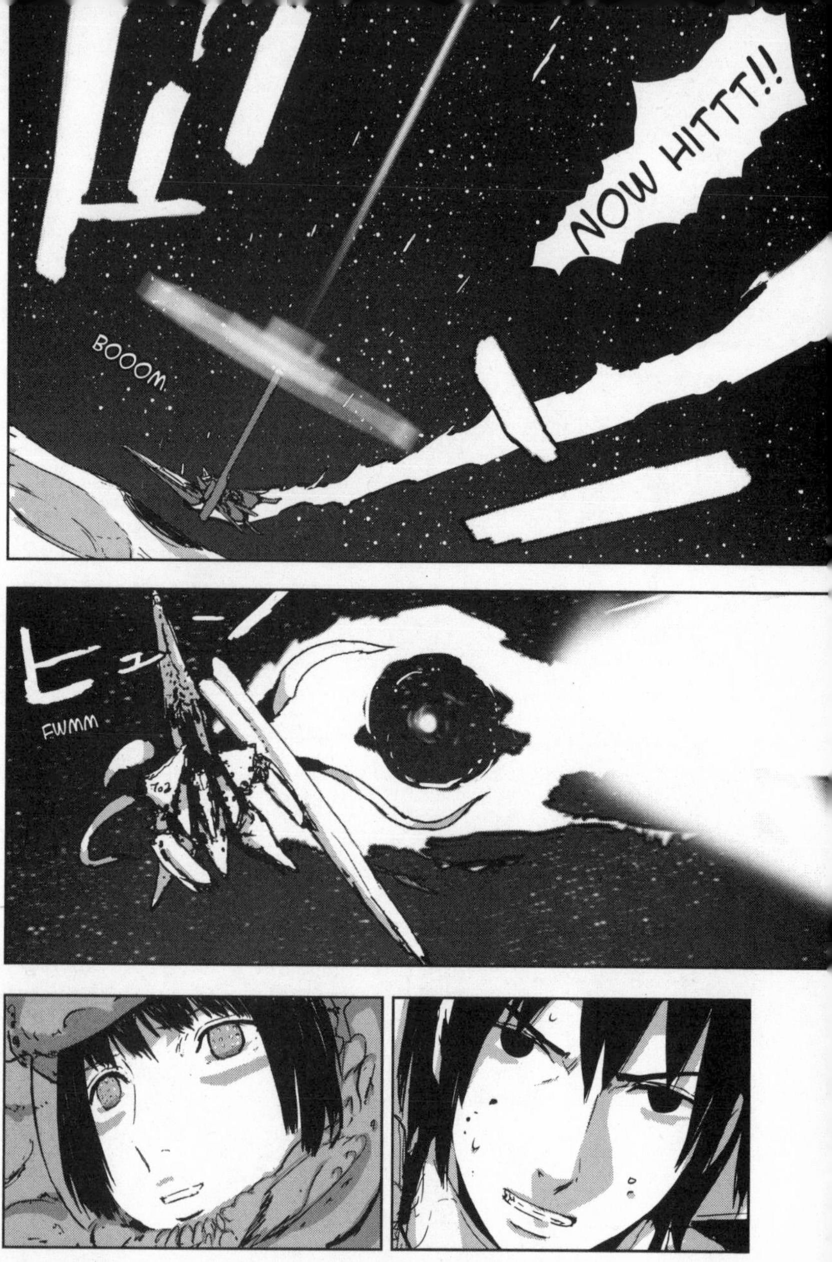
NOW HITTT!!
BOOOM.
ヒュー
FWMM

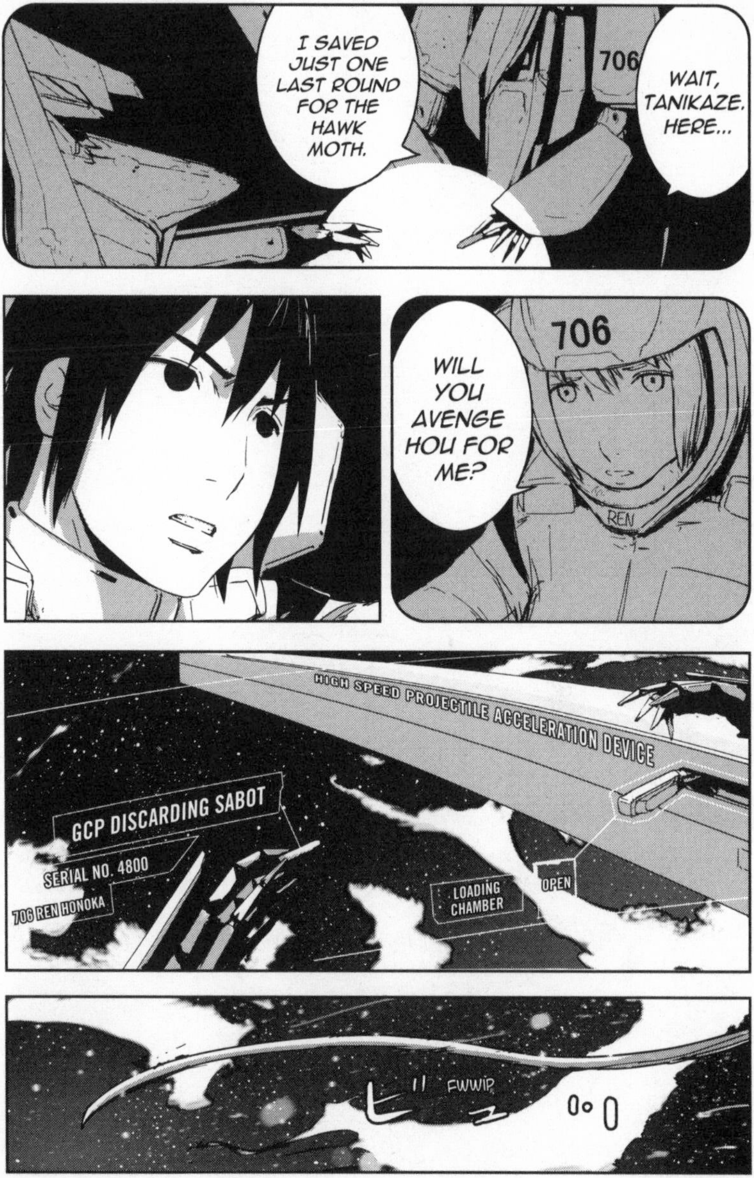
706
WAIT, TANIKAZE. HERE...
I SAVED JUST ONE LAST ROUND FOR THE HAWK MOTH.
706
WILL YOU AVENGE HOU FOR ME?
REN
HIGH SPEED PROJECTILE ACCELERATION DEVICE
OPEN
LOADING CHAMBER
GCP DISCARDING SABOT
SERIAL NO. 4800
706 REN HONOKA
FWWIP

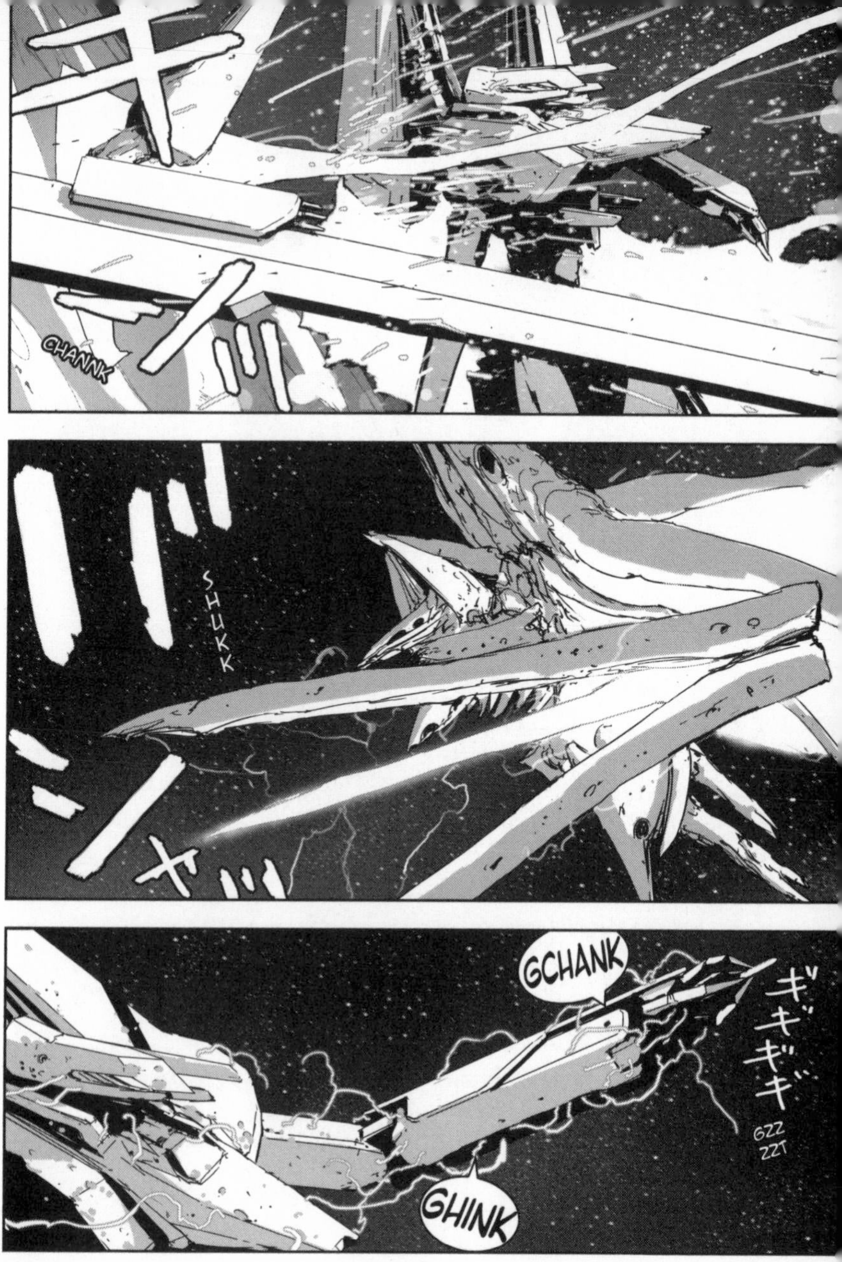
CHANNK
SHUKK
ギギギギ
GCHANK
GZZ ZZT
GHINK

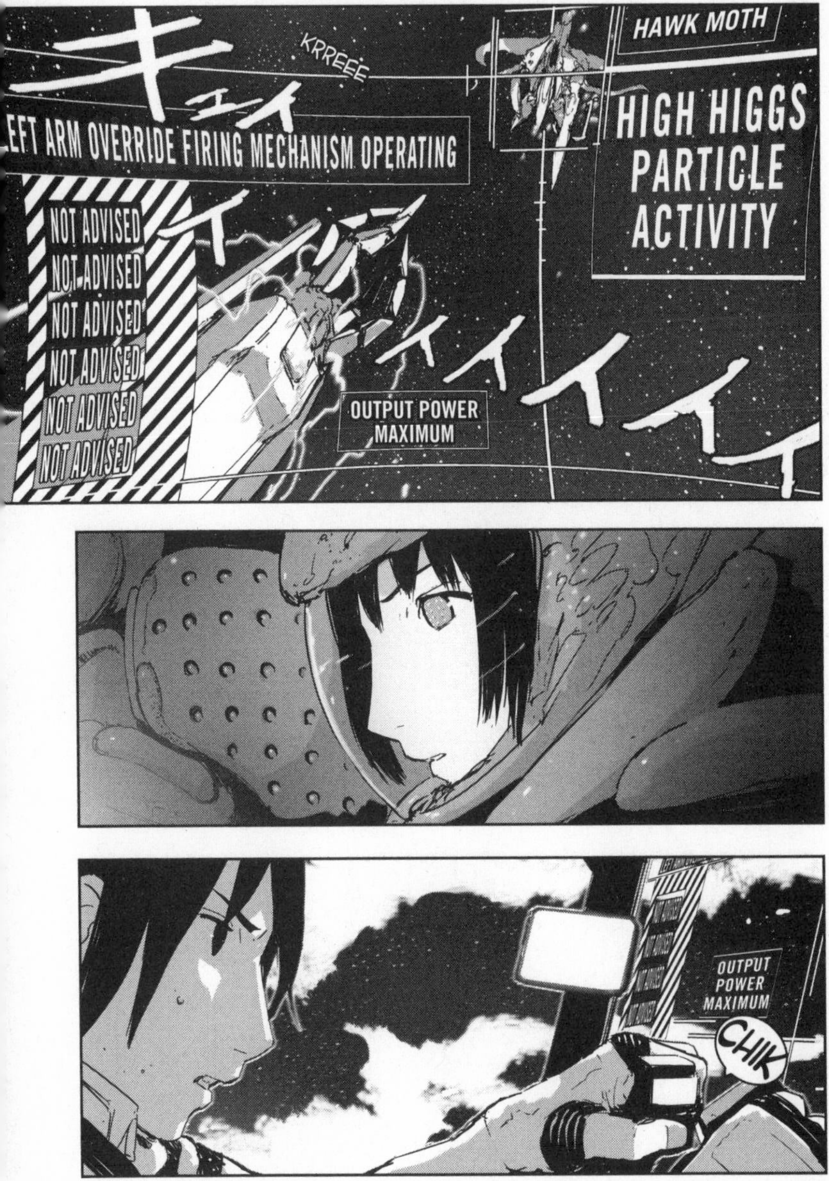
HAWK MOTH
HIGH HIGGS PARTICLE ACTIVITY
KRREEE
EFT ARM OVERRIDE FIRING MECHANISM OPERATING
NOT ADVISED
NOT ADVISED
NOT ADVISED
NOT ADVISED
NOT ADVISED
NOT ADVISED
OUTPUT POWER MAXIMUM
OUTPUT POWER MAXIMUM
CHIK

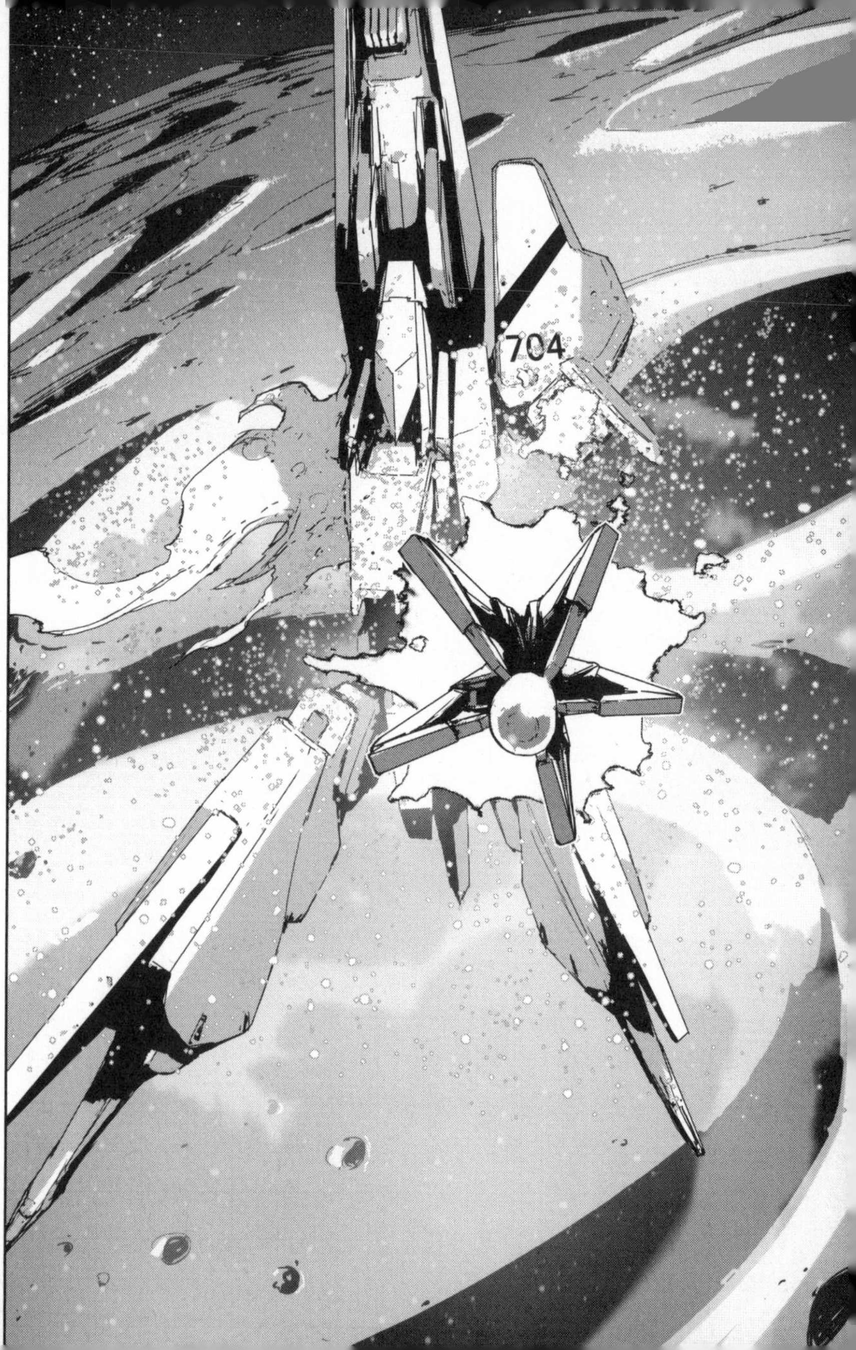
704

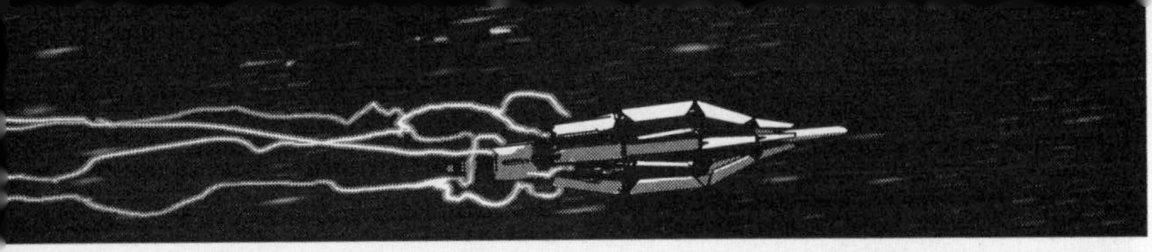

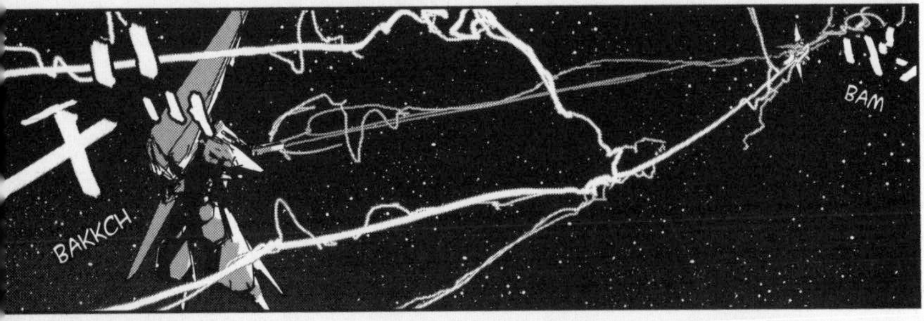
BAM
BAKKCH

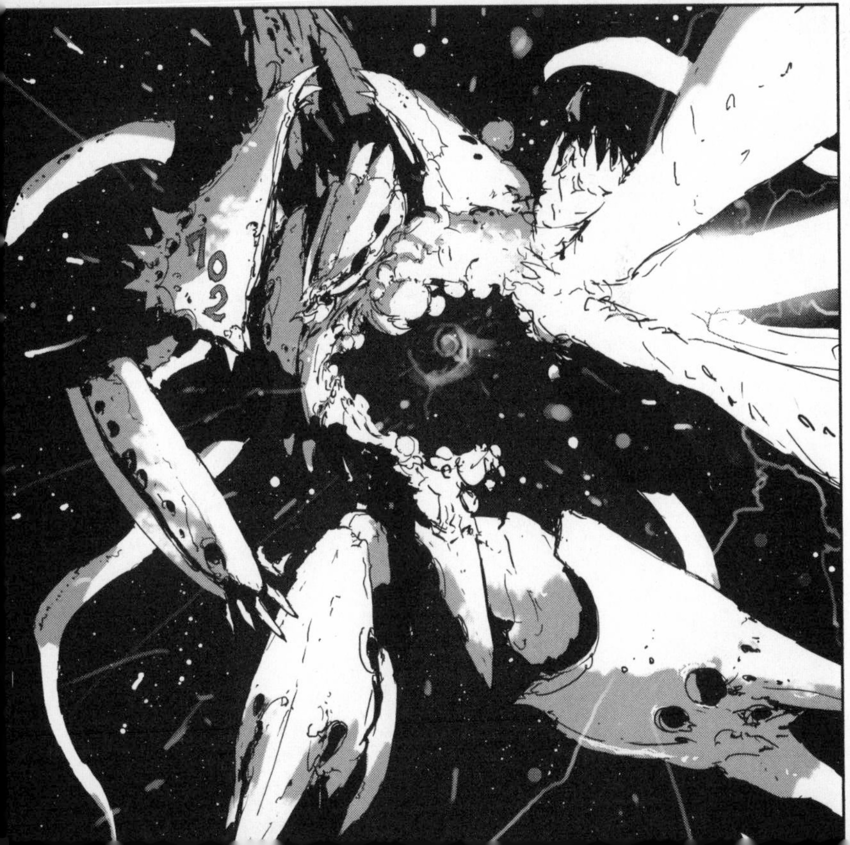
102

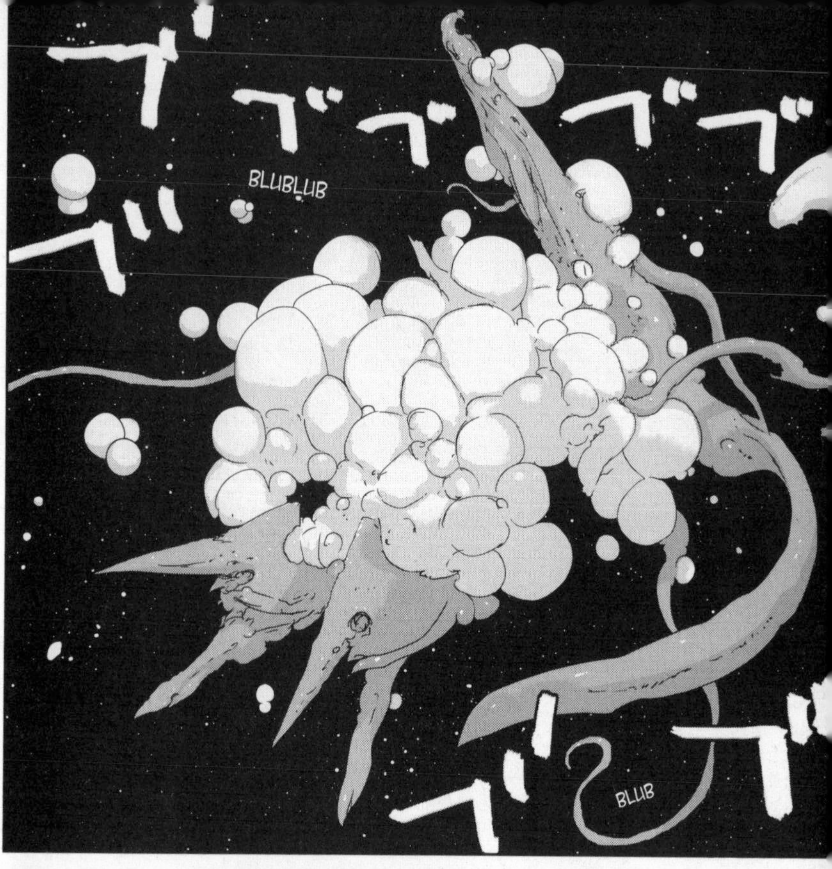
BLUBLUB
BLUB

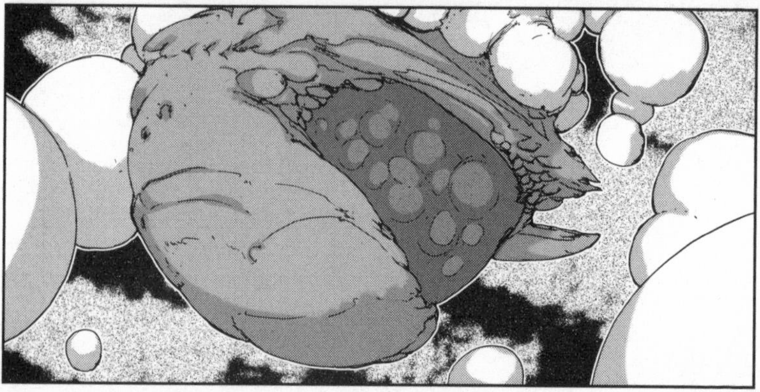

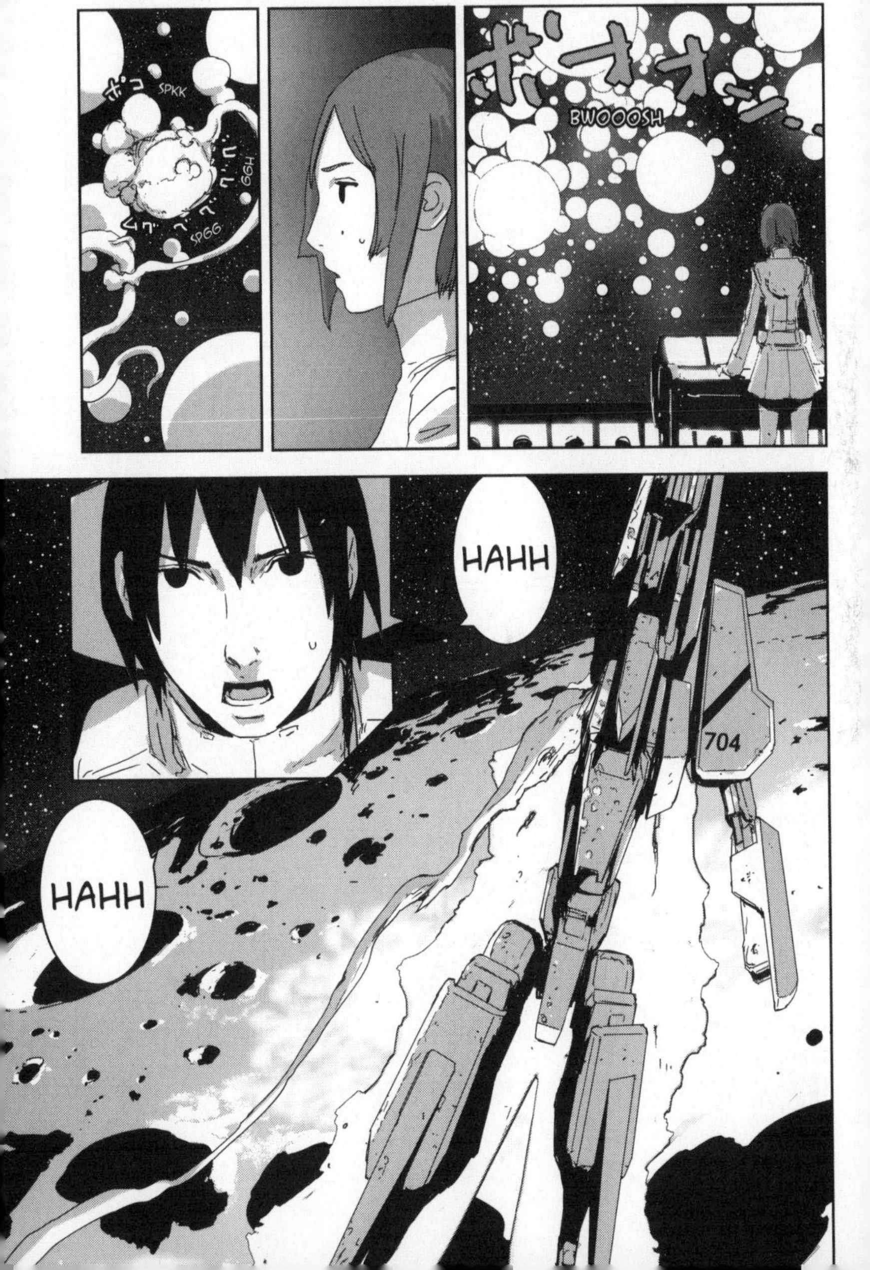
BWOOOSH
ポコ
SPKK
GGH
SPGG
HAHH
HAHH
704

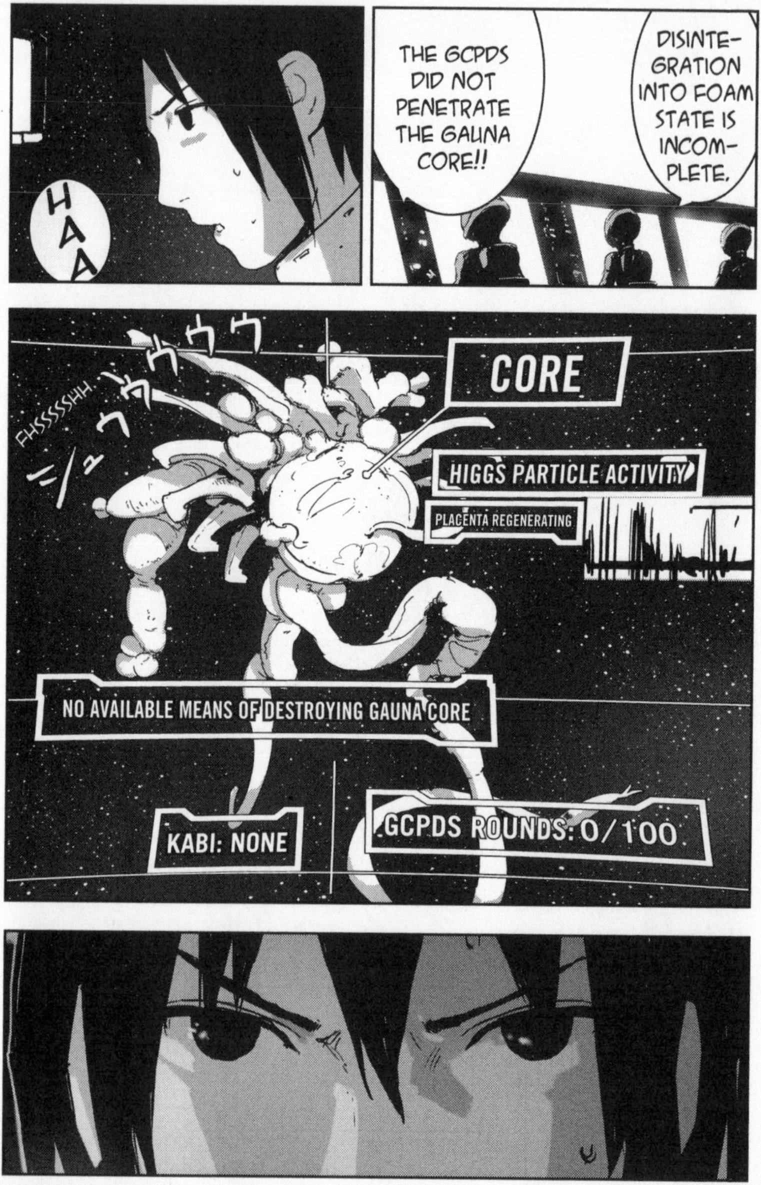
DISINTEGRATION INTO FOAM STATE IS INCOMPLETE.
THE GCPDS DID NOT PENETRATE THE GAUNA CORE!!
HAA
FHSSSSSHH
CORE
HIGGS PARTICLE ACTIVITY
PLACENTA REGENERATING
NO AVAILABLE MEANS OF DESTROYING GAUNA CORE
GCPDS ROUNDS: 0/100
KABI: NONE

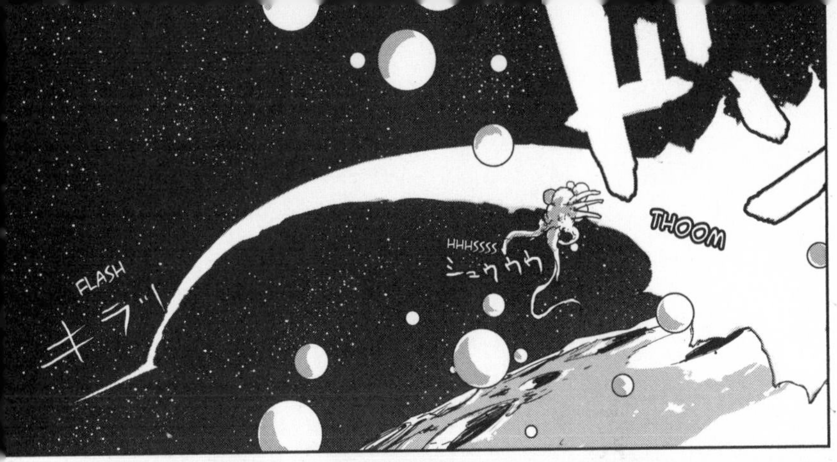
THOOM
HHHSSSS
FLASH
キラッ

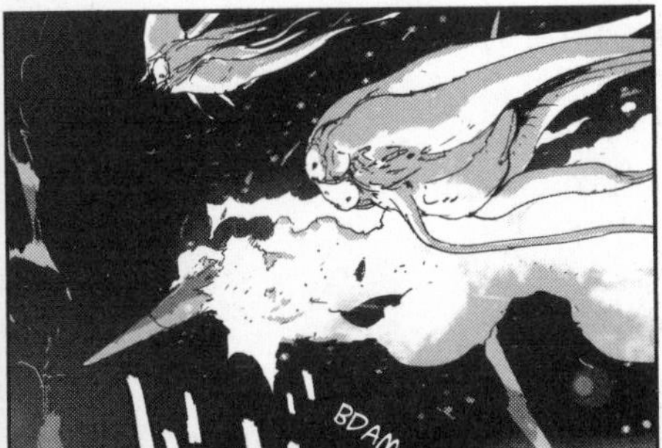
BDAM

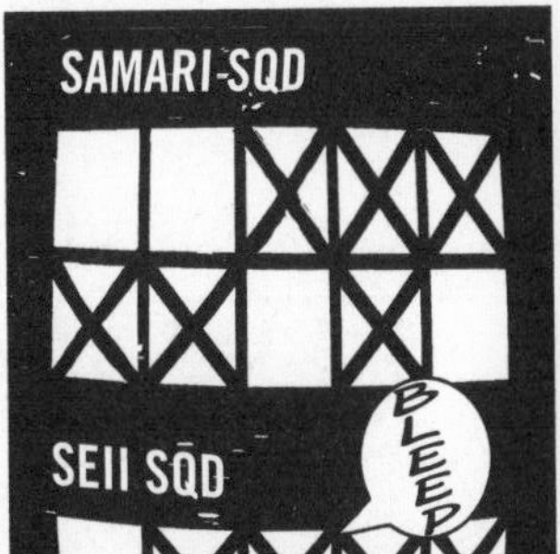
SAMARI-SQD
BLEEP
SEII SQD

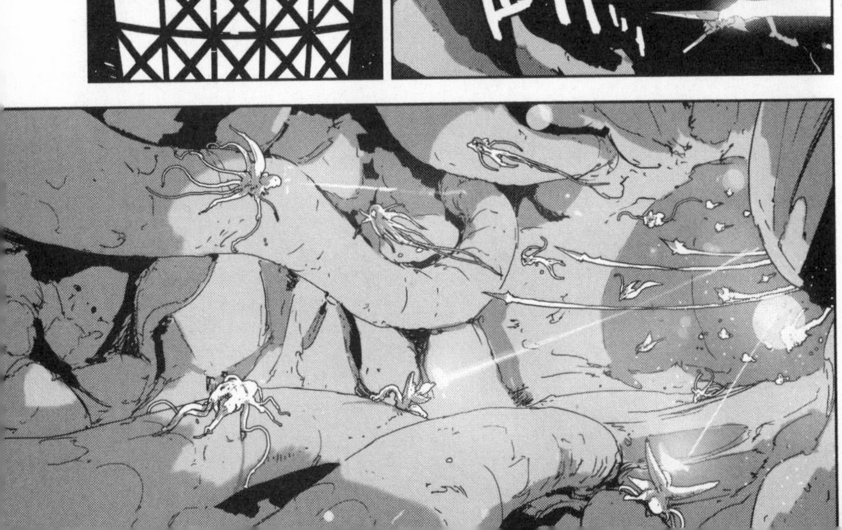

ASTEROID PROPULSION SOURCE
SAMARI - SEII SQD
IT'S IN SIGHT!!
THE ASTEROID'S PROPULSION SOURCE!!
WHRR
WHRR
WHRR
DOMDOMDOM
ZLRRR

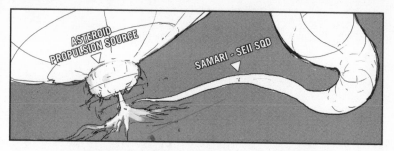

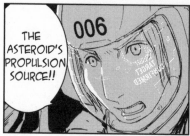

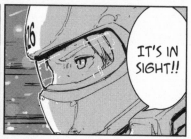

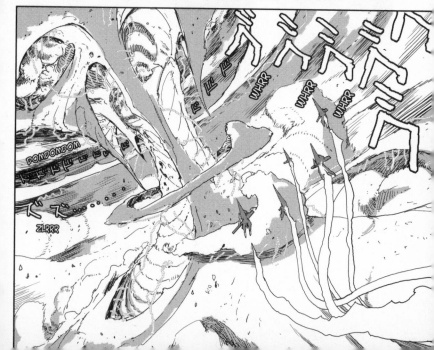

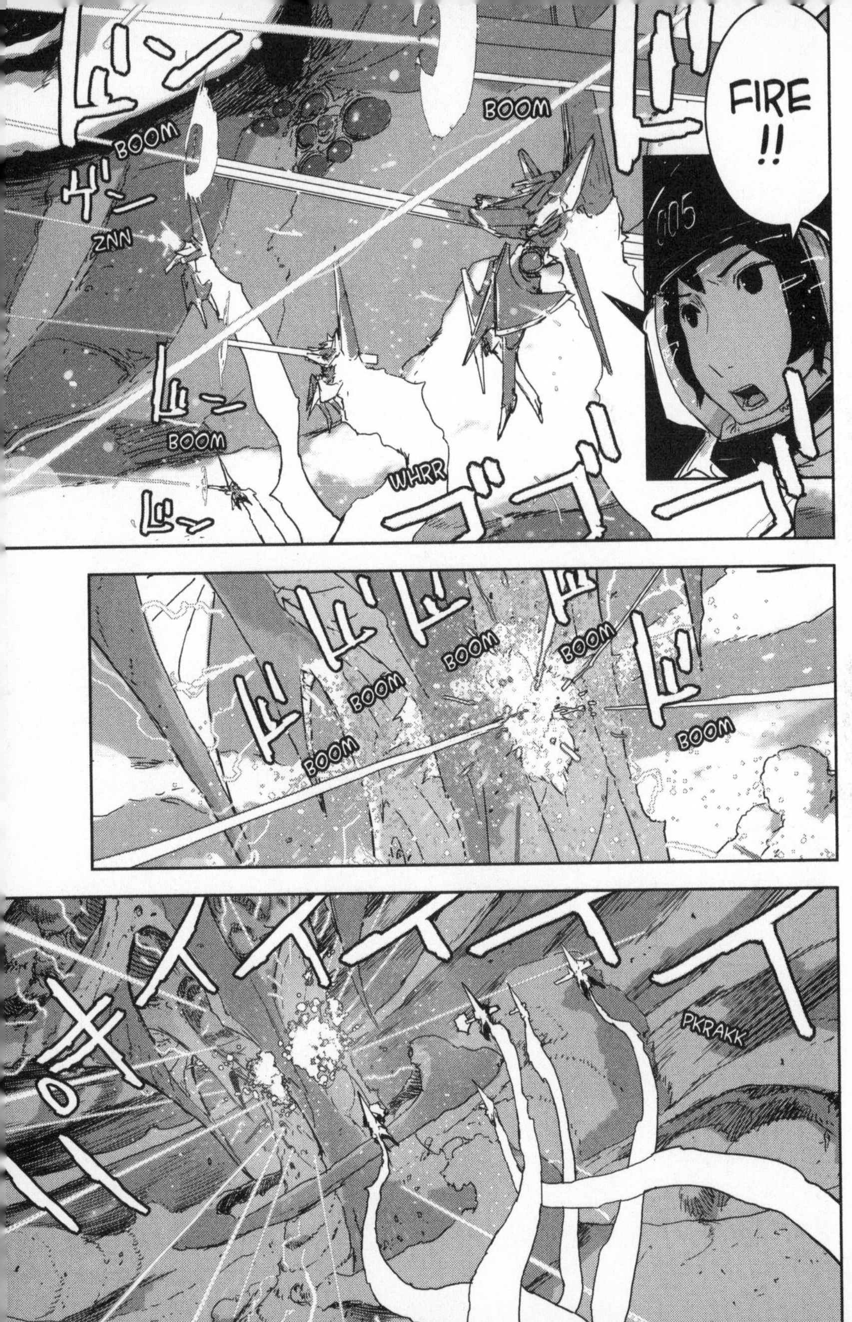
FIRE !!
005
BOOM
BOOM
ZNN
BOOM
WHRR
BOOM
BOOM
BOOM
BOOM
BOOM
PKRAKK

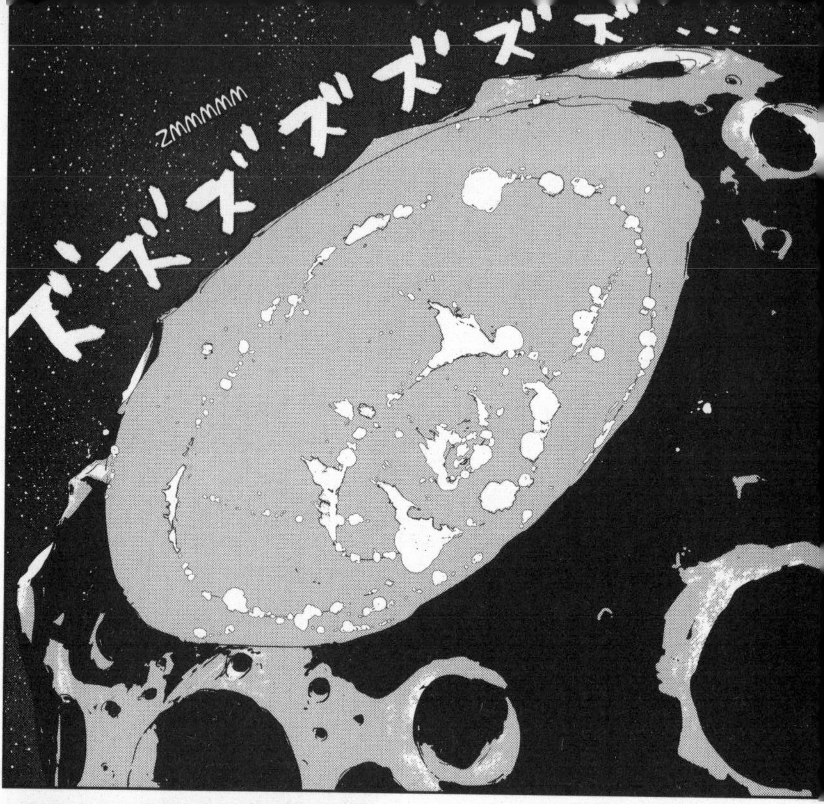
ZMMMMM

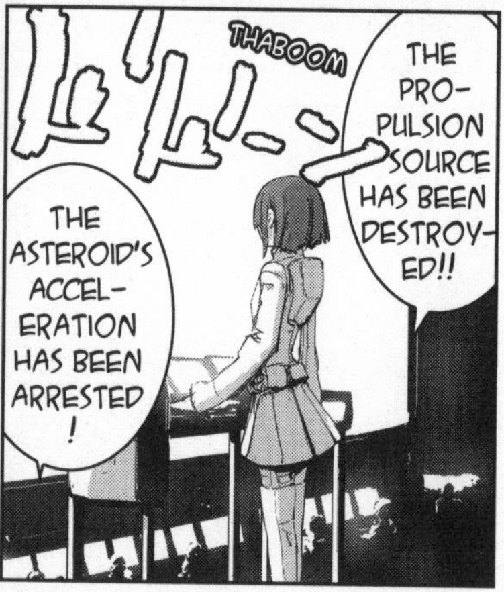
THABOOM
THE PRO-PULSION SOURCE HAS BEEN DESTROY-ED!!
THE ASTEROID'S ACCEL-ERATION HAS BEEN ARRESTED!

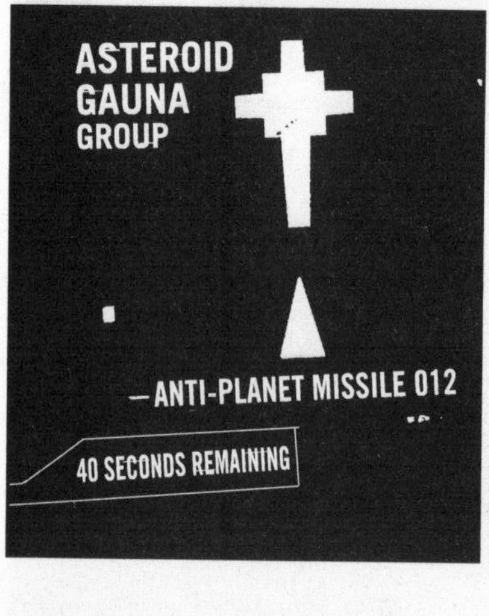
ASTEROID GAUNA GROUP
—ANTI-PLANET MISSILE 012
40 SECONDS REMAINING

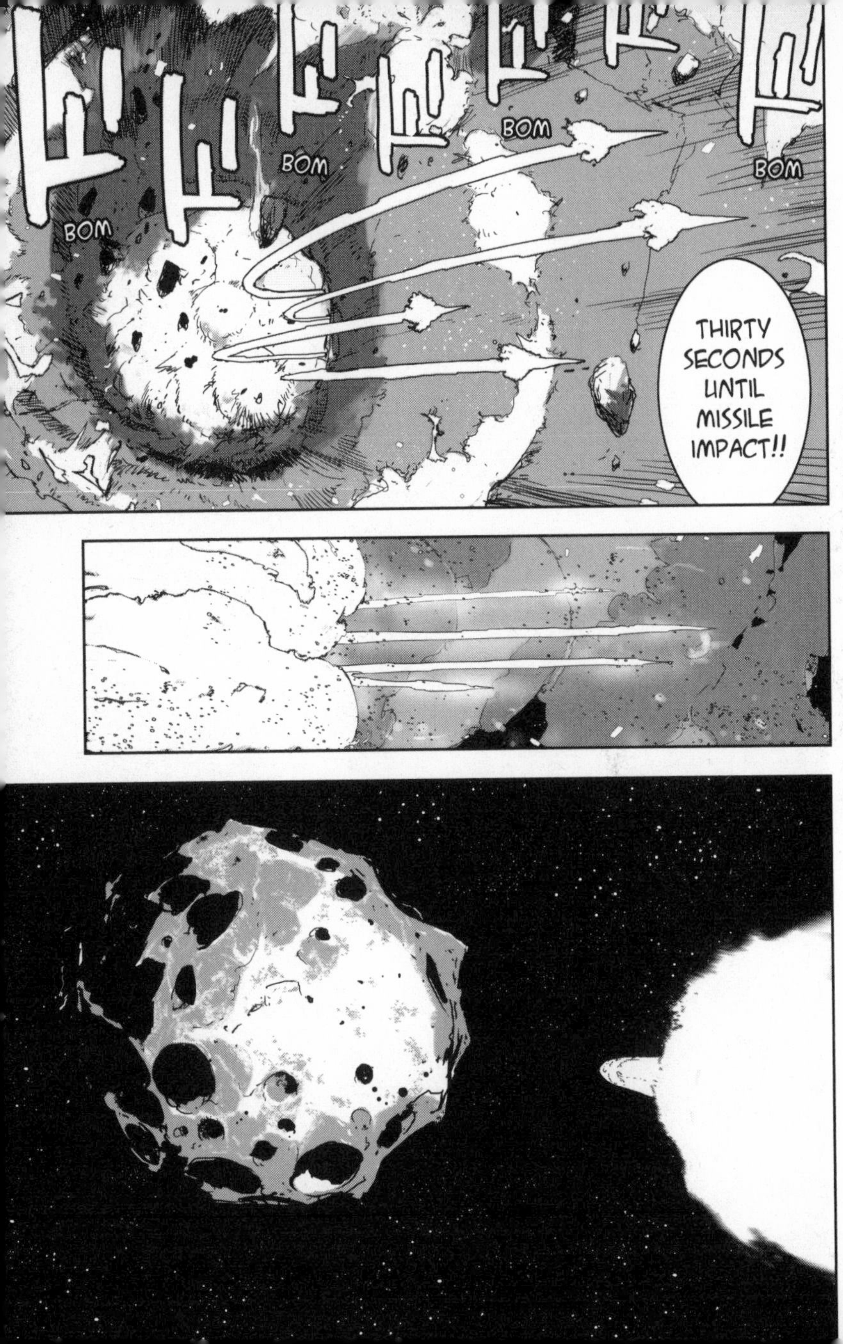
BOM
BOM
BOM
BOM
THIRTY SECONDS UNTIL MISSILE IMPACT!!

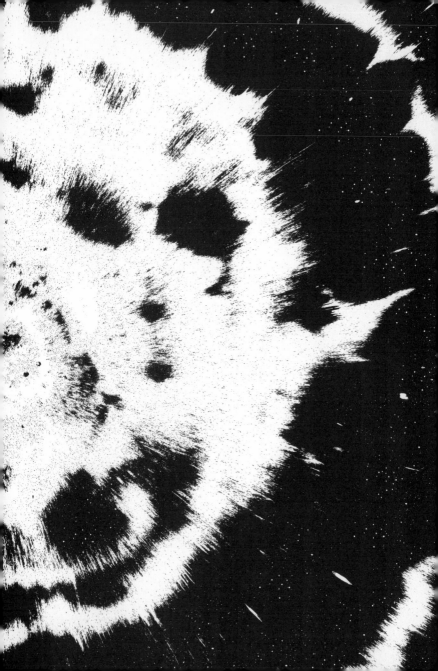

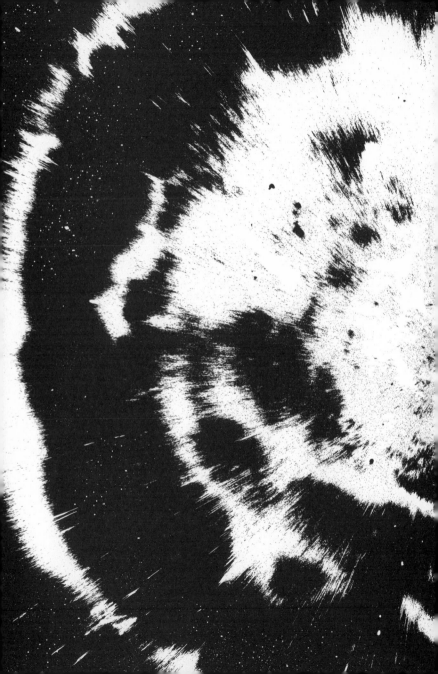

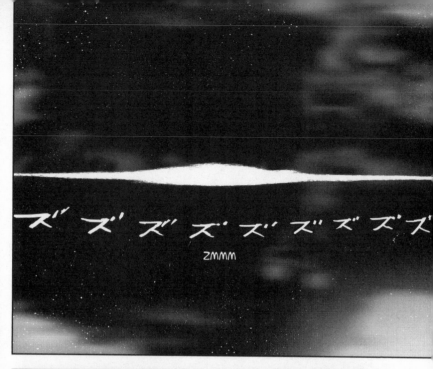

ズ ズ ズ ズ ズ ズ ズ ズ ズ

ZMMM

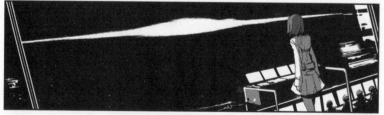

KRRKRR

ASTEROID
ELIMINATED
!!

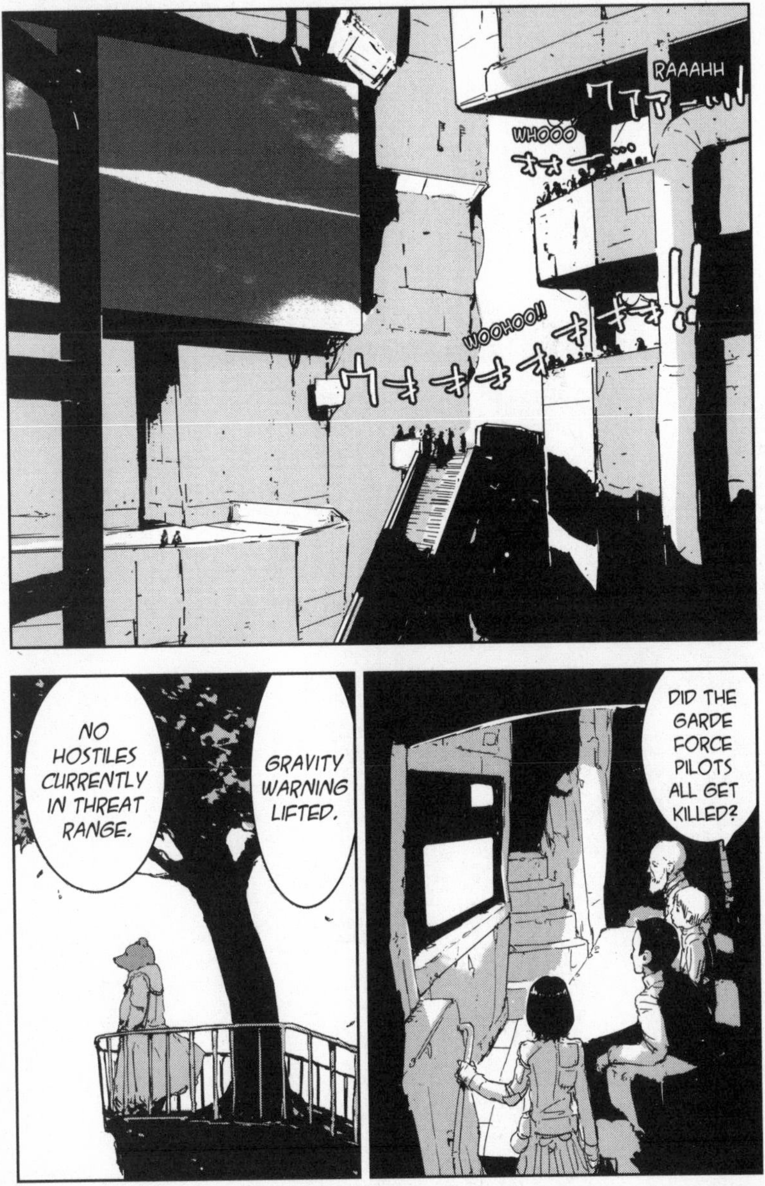
RAAAHH
WHOOO
WOOHOO!!
DID THE GARDE FORCE PILOTS ALL GET KILLED?
GRAVITY WARNING LIFTED.
NO HOSTILES CURRENTLY IN THREAT RANGE.

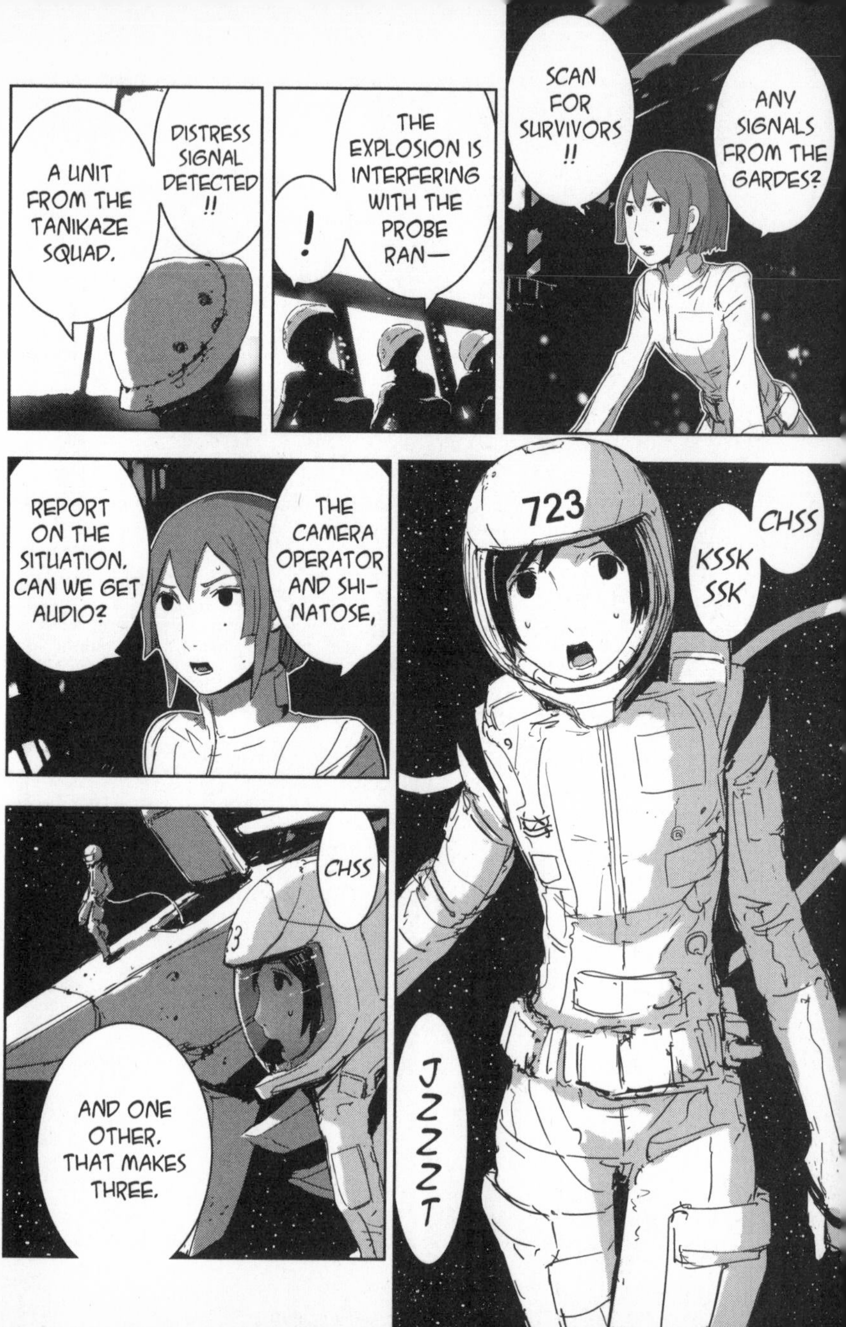
ANY SIGNALS FROM THE GARDES?
SCAN FOR SURVIVORS !!
THE EXPLOSION IS INTERFERING WITH THE PROBE RAN—
!
DISTRESS SIGNAL DETECTED !!
A UNIT FROM THE TANIKAZE SQUAD.
723
CHSS
KSSK SSK
JZZZT
THE CAMERA OPERATOR AND SHI-NATOSE,
REPORT ON THE SITUATION. CAN WE GET AUDIO?
CHSS
AND ONE OTHER. THAT MAKES THREE.

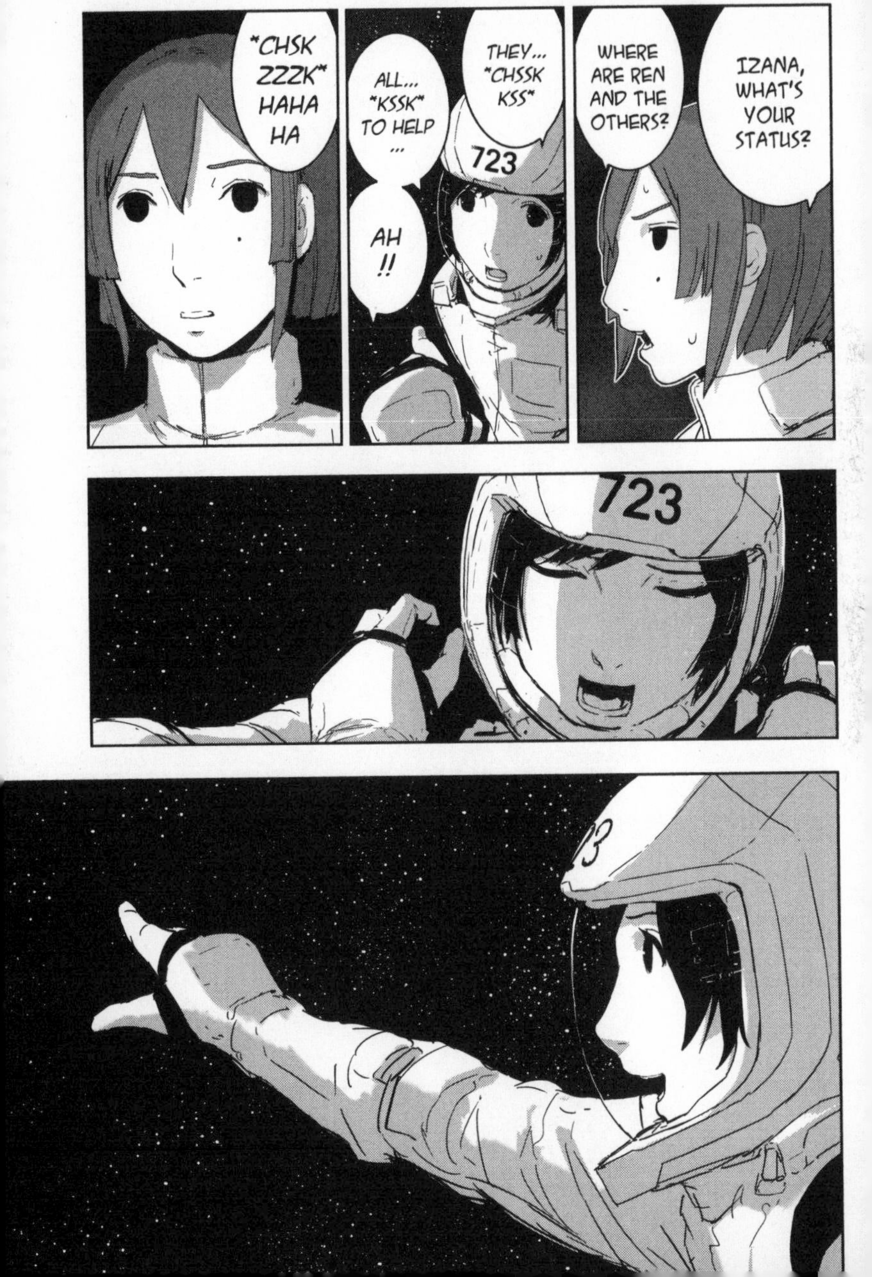
IZANA, WHAT'S YOUR STATUS?
WHERE ARE REN AND THE OTHERS?
THEY... *CHSSK KSS*
ALL... *KSSK* TO HELP ...
AH !!
CHSK ZZZK HAHA HA

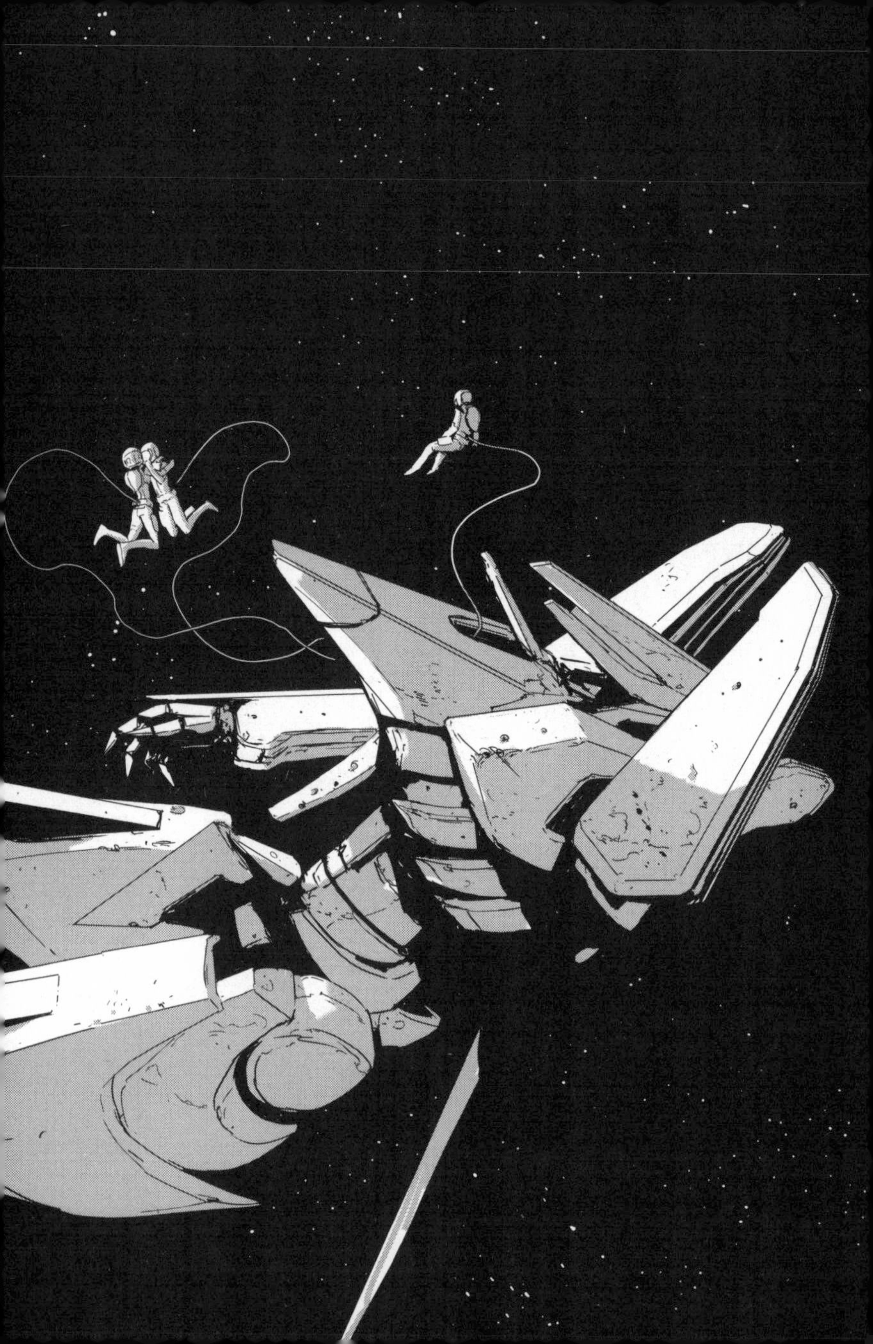

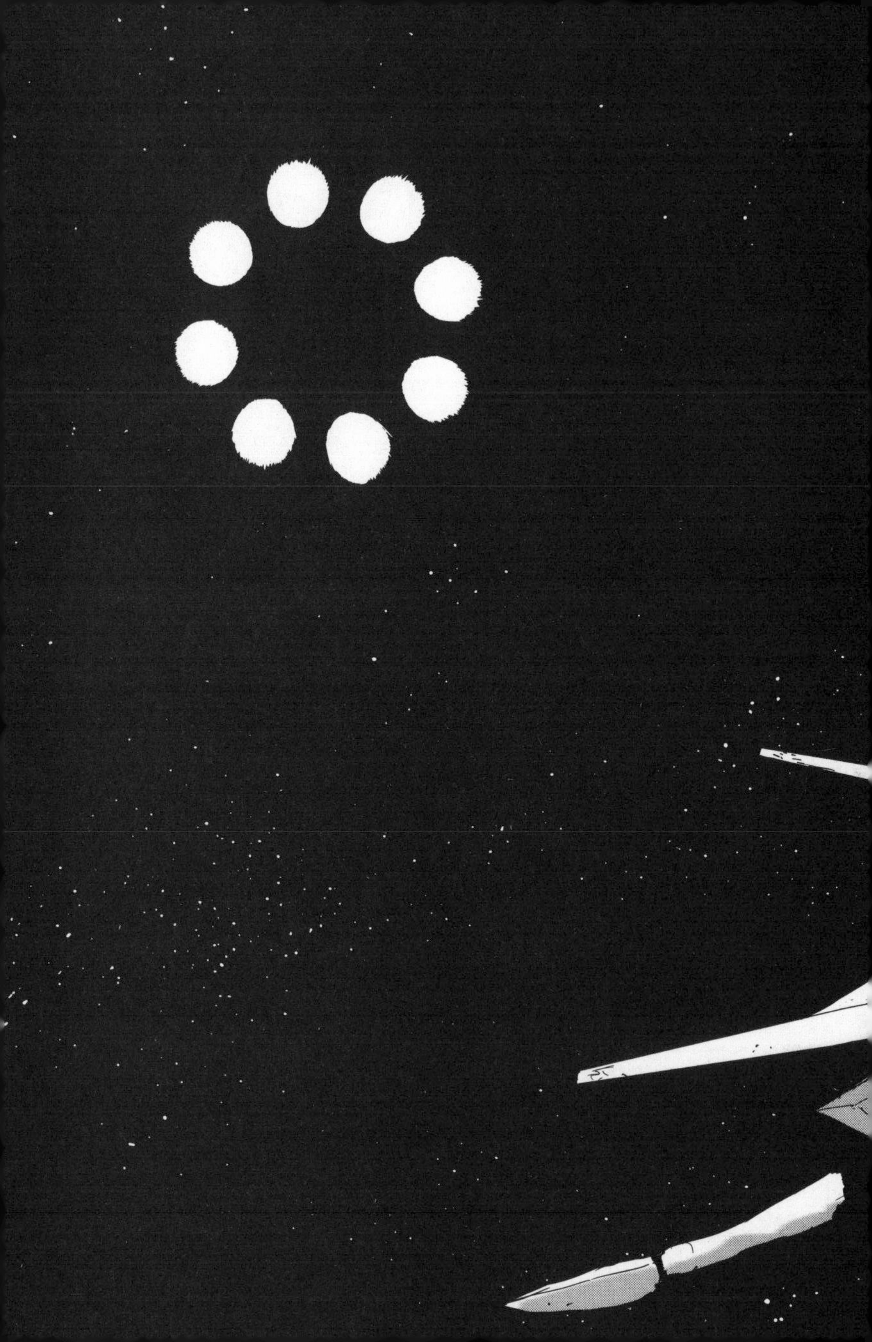

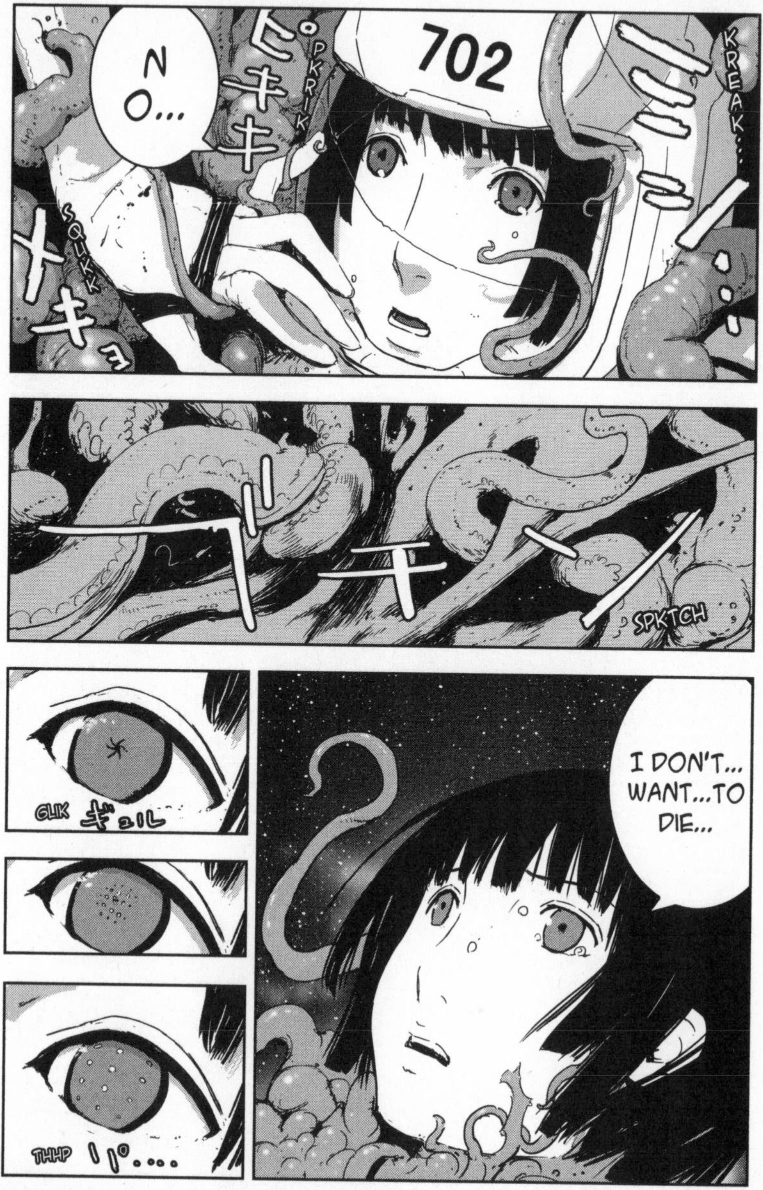
KREAK...
702
PKRIK
NO...
SQUKK
SPKTCH
I DON'T... WANT...TO DIE...
GLIK
THHP

THE HAWK MOTH'S CORE REGREW ITS LOST PLACENTA.

DURING THE PROCESS, THE PILOT-SHAPED PORTION RELIVED THE DEATH OF ITS HUMAN PREDECESSOR FOR THE BRIEFEST OF MOMENTS.

THE MEMORY ALWAYS VANISHED QUICKLY, BUT A SURE FEELING OF DEJA VU REMAINED.

THE SENSATION ACCUMULATED AND INTENSIFIED WITH EACH RECONSTITUTION...

Chapter 26: END

Chapter 27: Toha Heavy Industries' Resumption

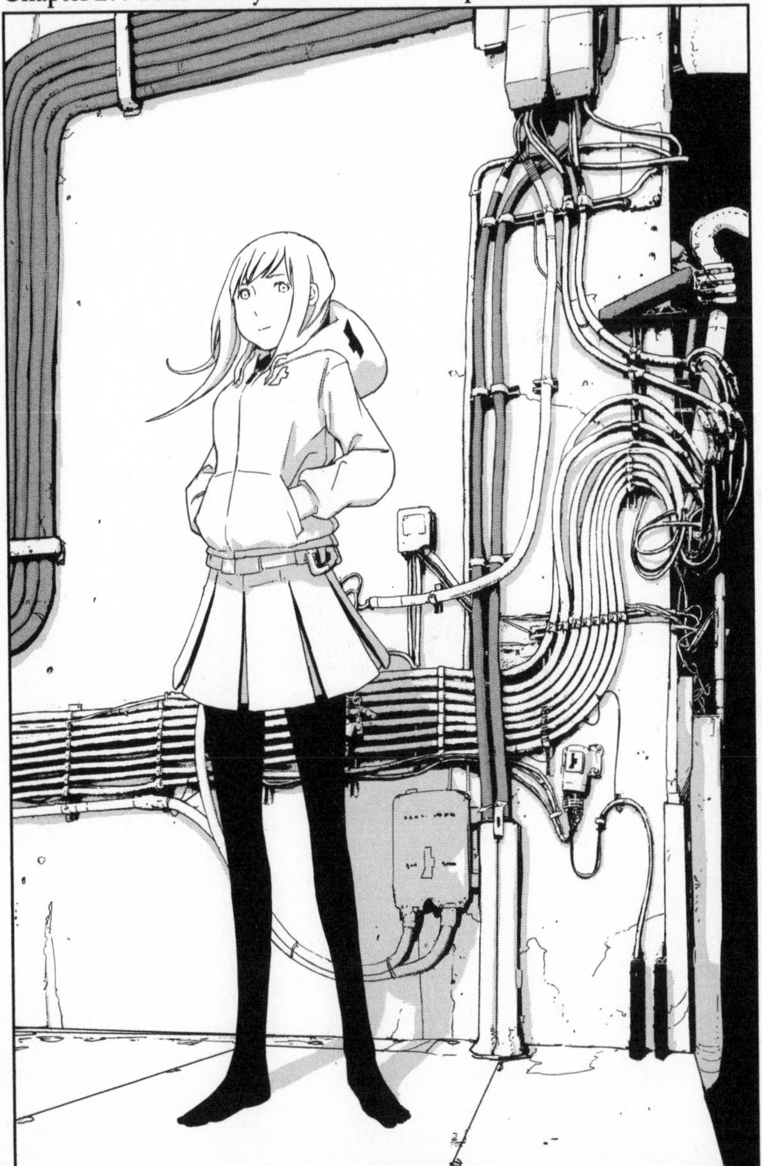

One Hundred Sights of Sidonia Part Twenty-One:
Sidonia Hospital No. 3 Rooftop

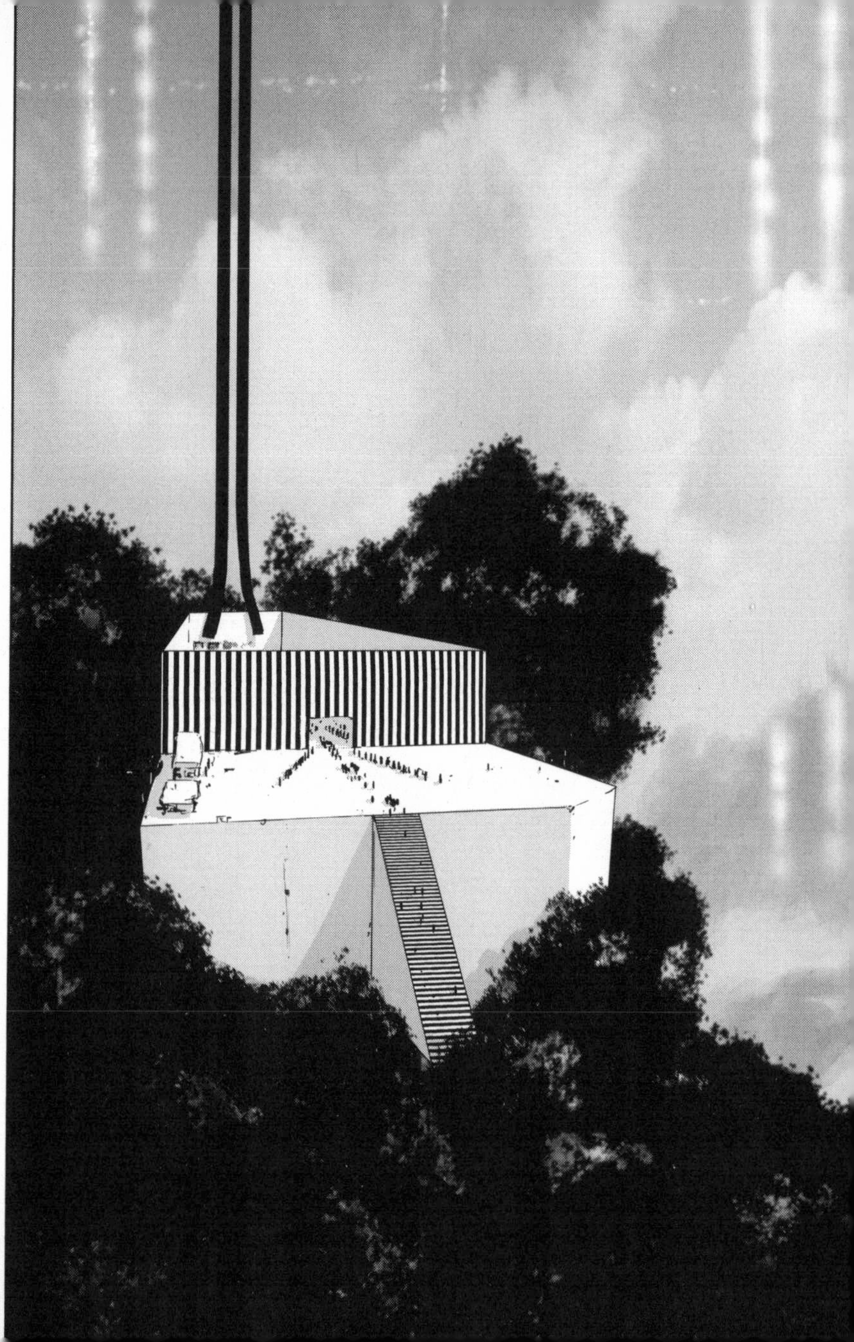

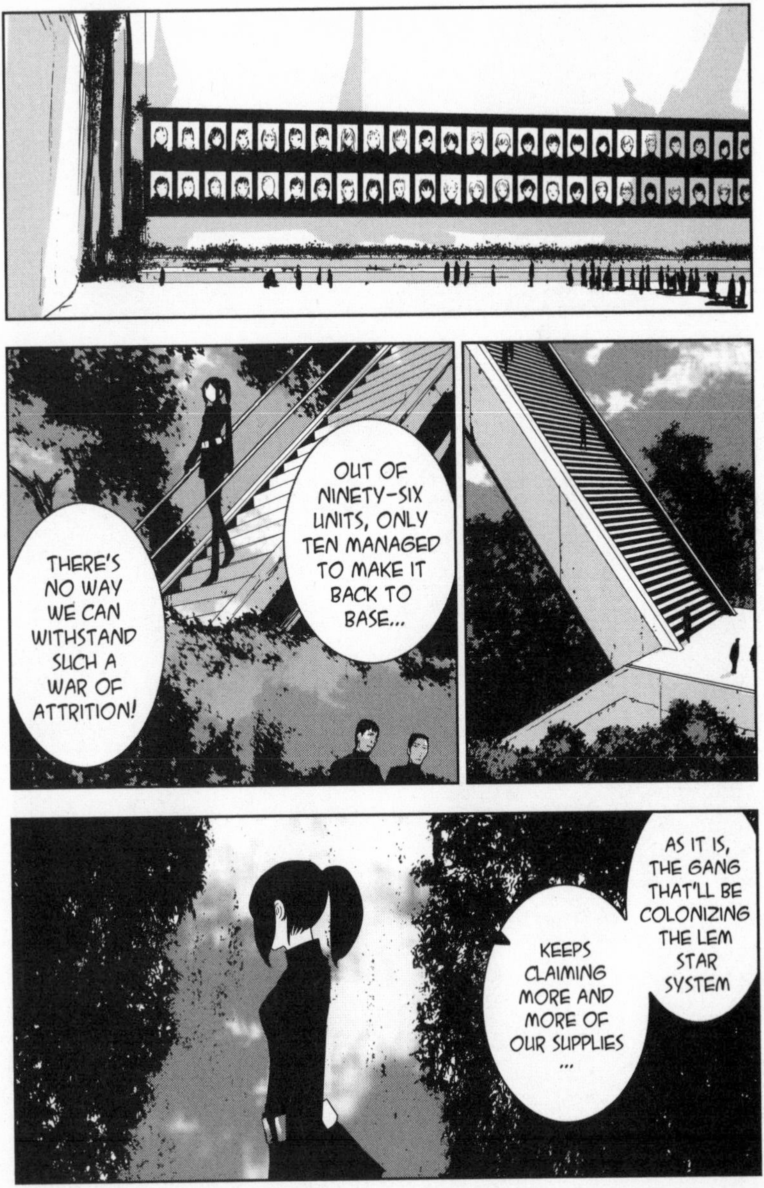
OUT OF NINETY-SIX UNITS, ONLY TEN MANAGED TO MAKE IT BACK TO BASE...
THERE'S NO WAY WE CAN WITHSTAND SUCH A WAR OF ATTRITION!
AS IT IS, THE GANG THAT'LL BE COLONIZING THE LEM STAR SYSTEM
KEEPS CLAIMING MORE AND MORE OF OUR SUPPLIES ...

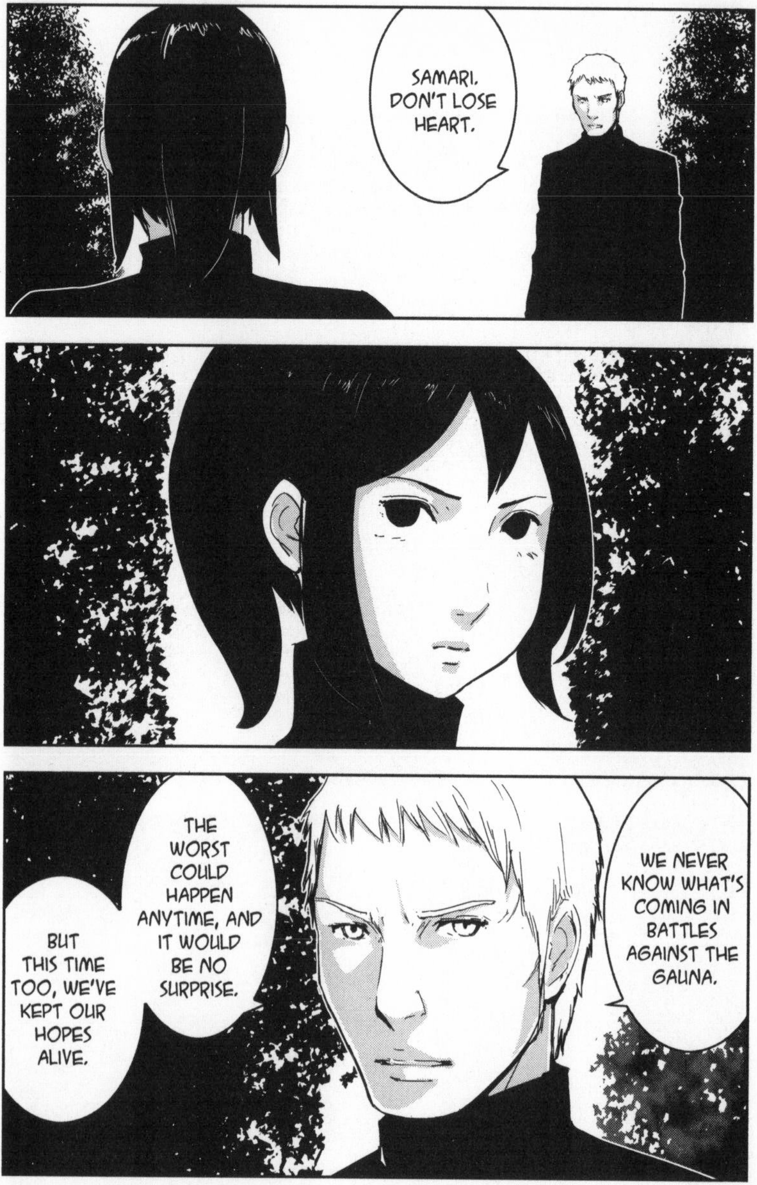
SAMARI. DON'T LOSE HEART.
WE NEVER KNOW WHAT'S COMING IN BATTLES AGAINST THE GAUNA.
THE WORST COULD HAPPEN ANYTIME, AND IT WOULD BE NO SURPRISE.
BUT THIS TIME TOO, WE'VE KEPT OUR HOPES ALIVE.

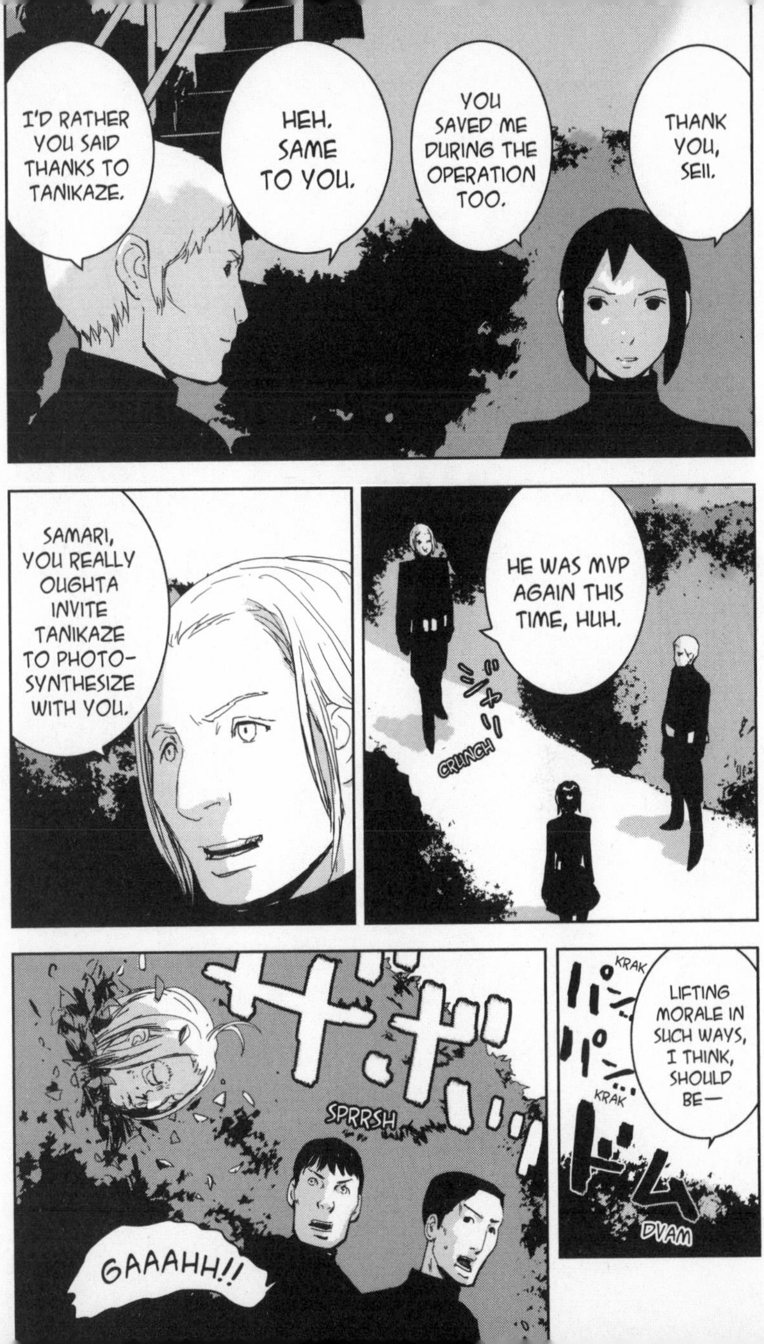
THANK YOU, SEII.
YOU SAVED ME DURING THE OPERATION TOO.
HEH. SAME TO YOU.
I'D RATHER YOU SAID THANKS TO TANIKAZE.
HE WAS MVP AGAIN THIS TIME, HUH.
CRUNCH
SAMARI, YOU REALLY OUGHTA INVITE TANIKAZE TO PHOTO-SYNTHESIZE WITH YOU.
KRAK
LIFTING MORALE IN SUCH WAYS, I THINK, SHOULD BE—
KRAK
DVAM
SPRRSH
GAAAHH!!

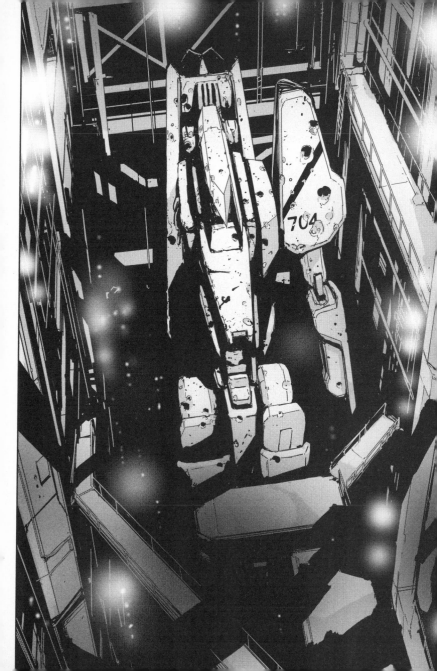

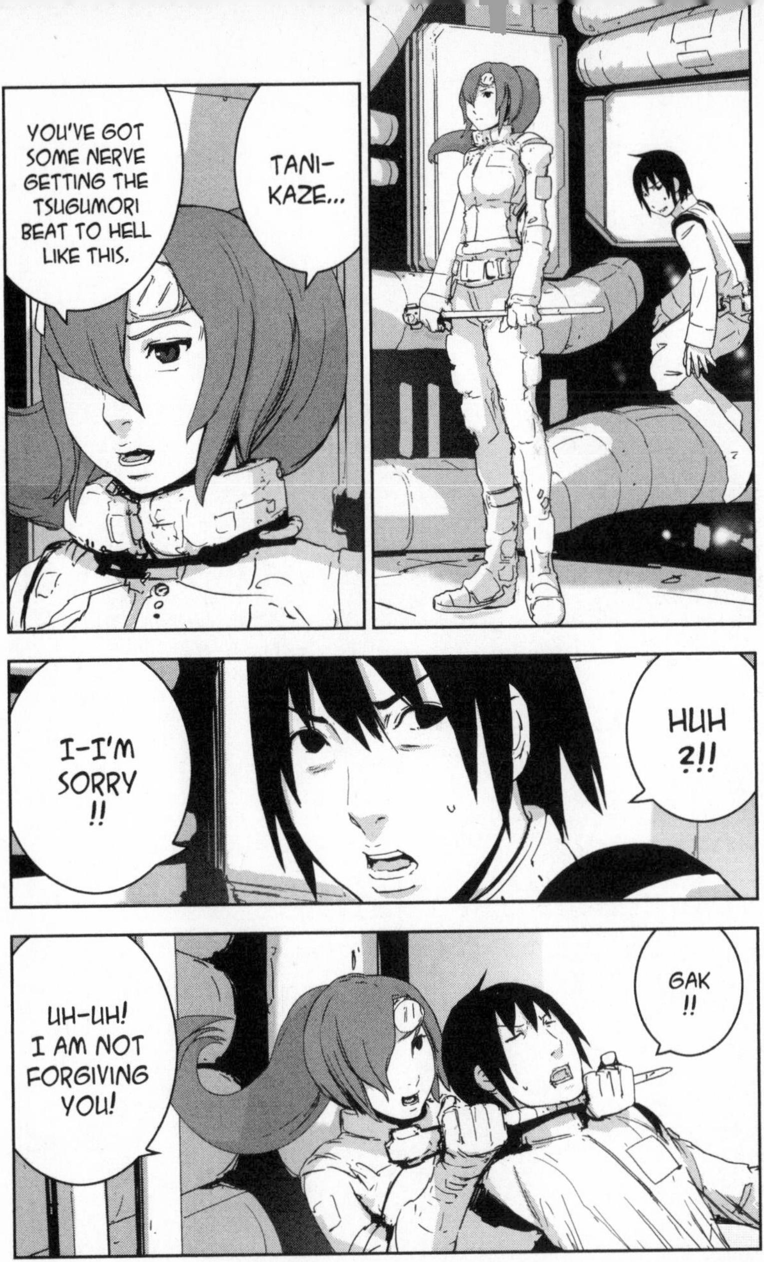
TANI-KAZE...
YOU'VE GOT SOME NERVE GETTING THE TSUGUMORI BEAT TO HELL LIKE THIS.
HUH ?!!
I-I'M SORRY !!
GAK !!
UH-UH! I AM NOT FORGIVING YOU!

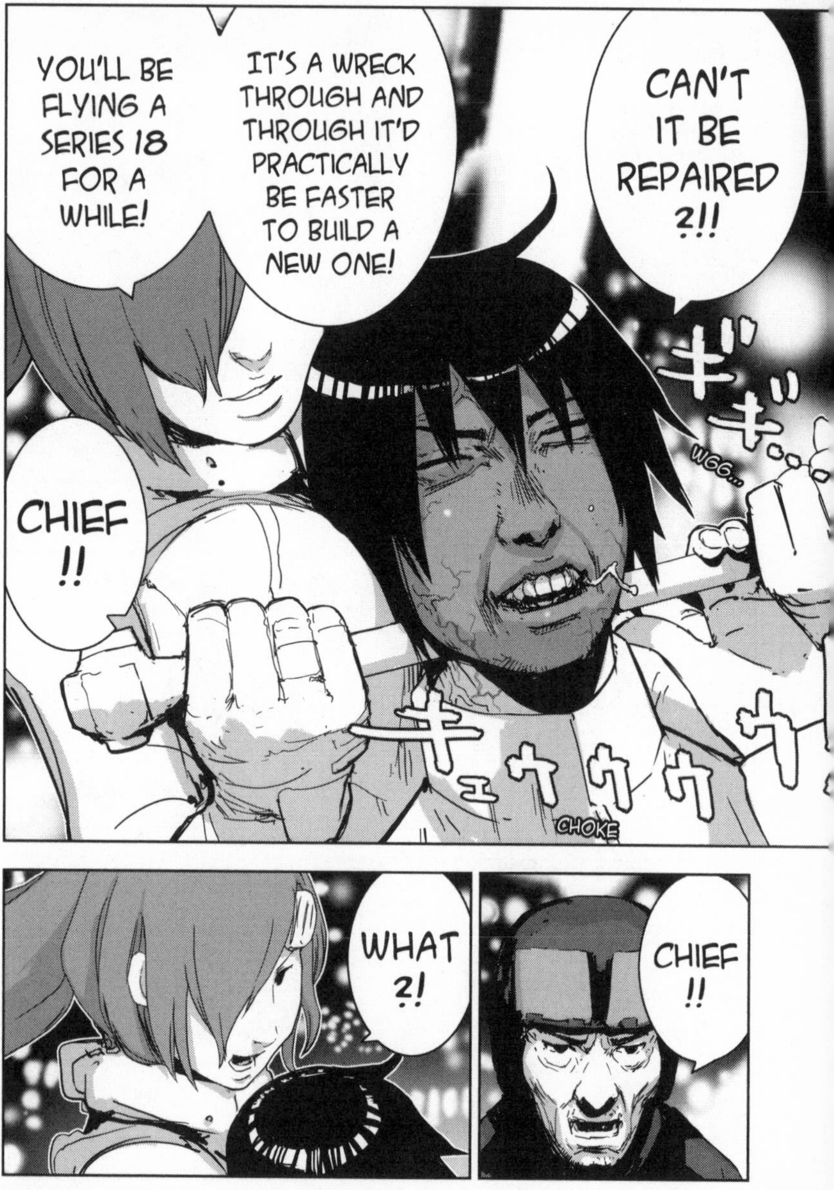
CAN'T IT BE REPAIRED ?!!
IT'S A WRECK THROUGH AND THROUGH IT'D PRACTICALLY BE FASTER TO BUILD A NEW ONE!
YOU'LL BE FLYING A SERIES 18 FOR A WHILE!
ギギ…
WGG...
CHIEF !!
キュウウウウ
CHOKE
CHIEF !!
WHAT ?!

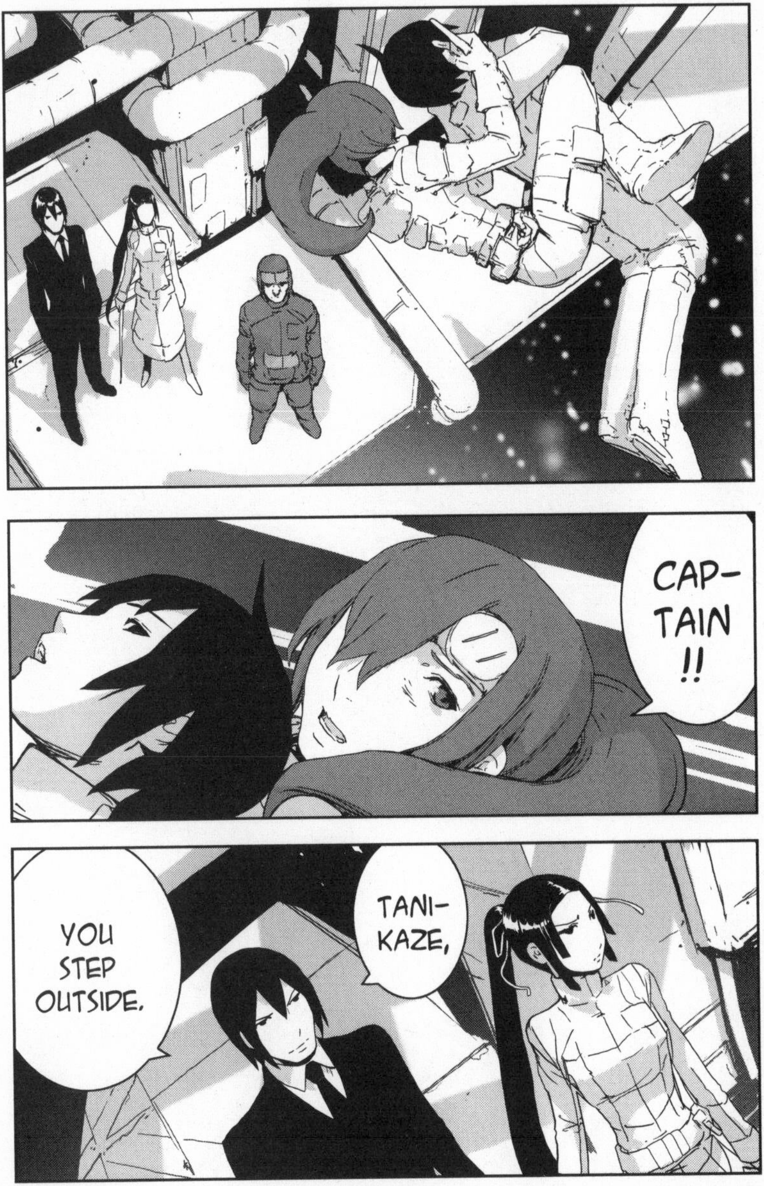
CAP-TAIN !!
TANI-KAZE,
YOU STEP OUTSIDE.

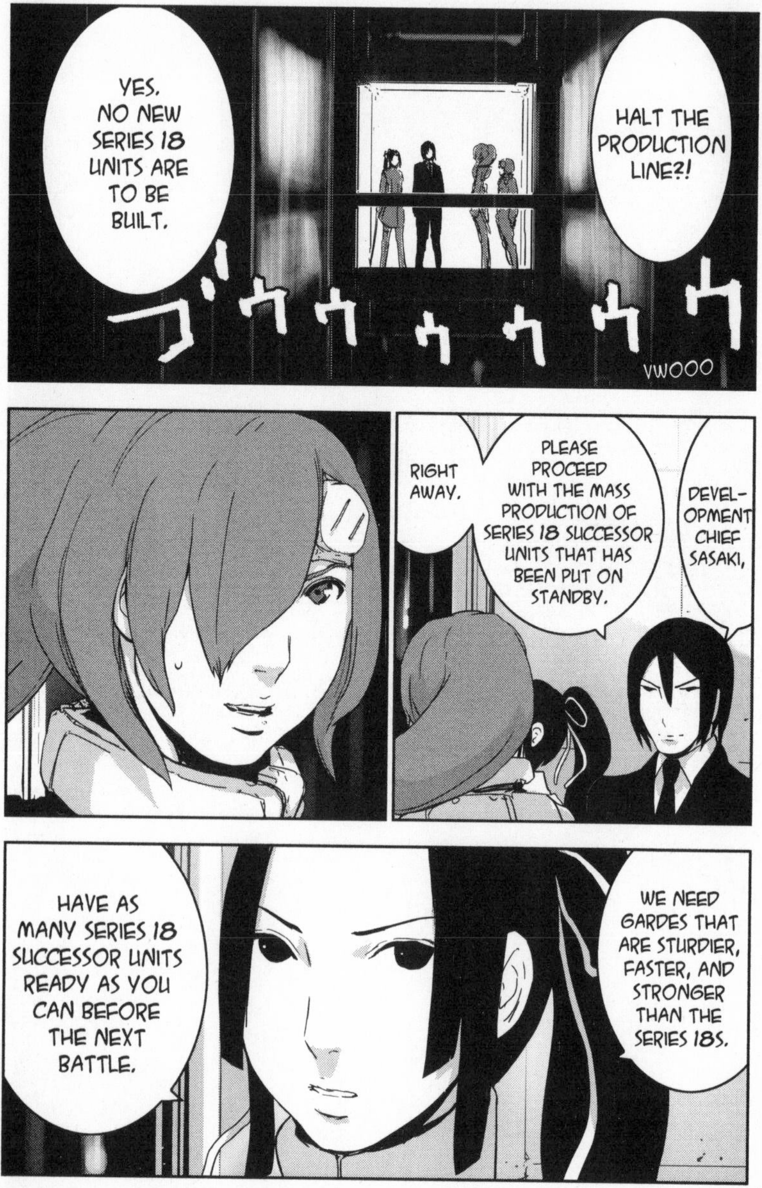
HALT THE PRODUCTION LINE?!
YES. NO NEW SERIES 18 UNITS ARE TO BE BUILT.
VWOOO
DEVEL-OPMENT CHIEF SASAKI,
PLEASE PROCEED WITH THE MASS PRODUCTION OF SERIES 18 SUCCESSOR UNITS THAT HAS BEEN PUT ON STANDBY.
RIGHT AWAY.
WE NEED GARDES THAT ARE STURDIER, FASTER, AND STRONGER THAN THE SERIES 18S.
HAVE AS MANY SERIES 18 SUCCESSOR UNITS READY AS YOU CAN BEFORE THE NEXT BATTLE.

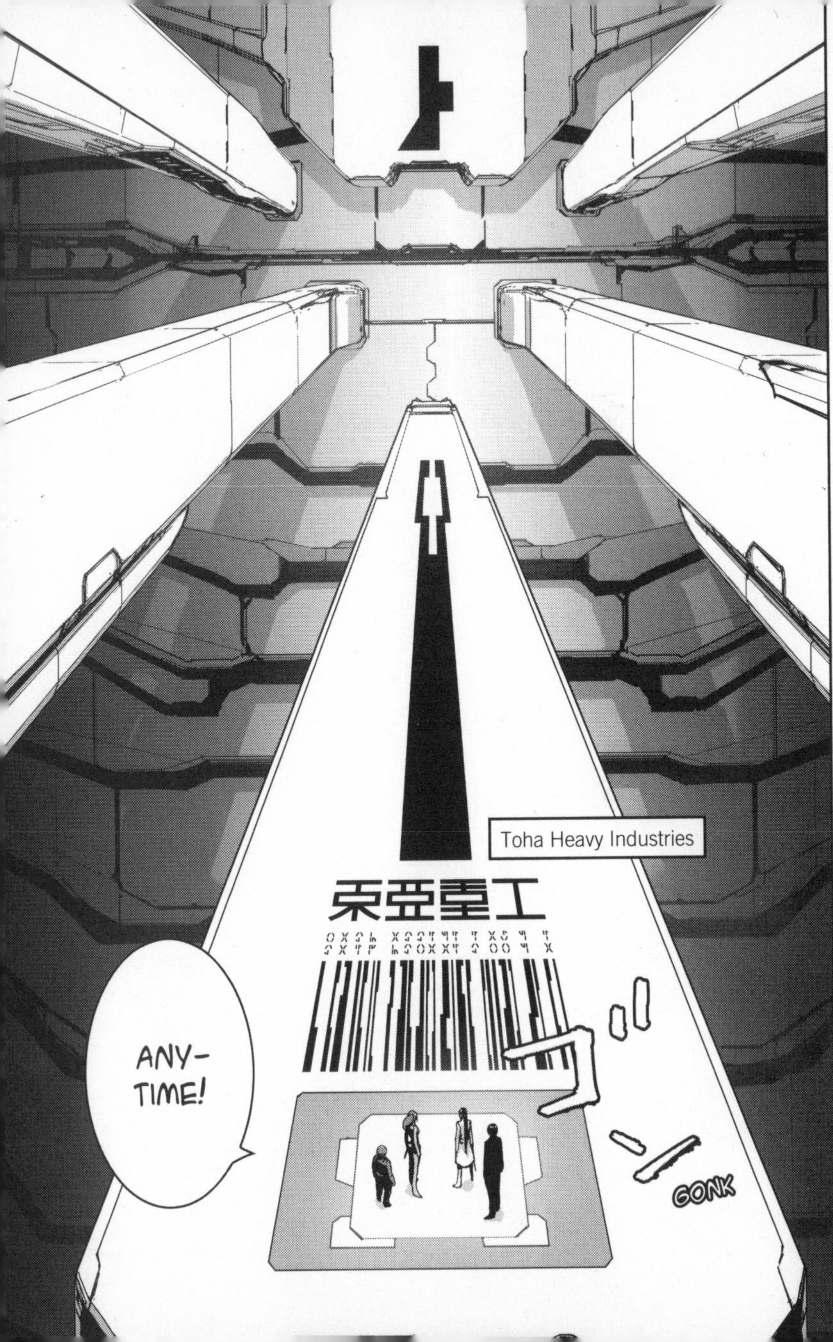
Toha Heavy Industries
東亜重工
GONK
ANY-TIME!

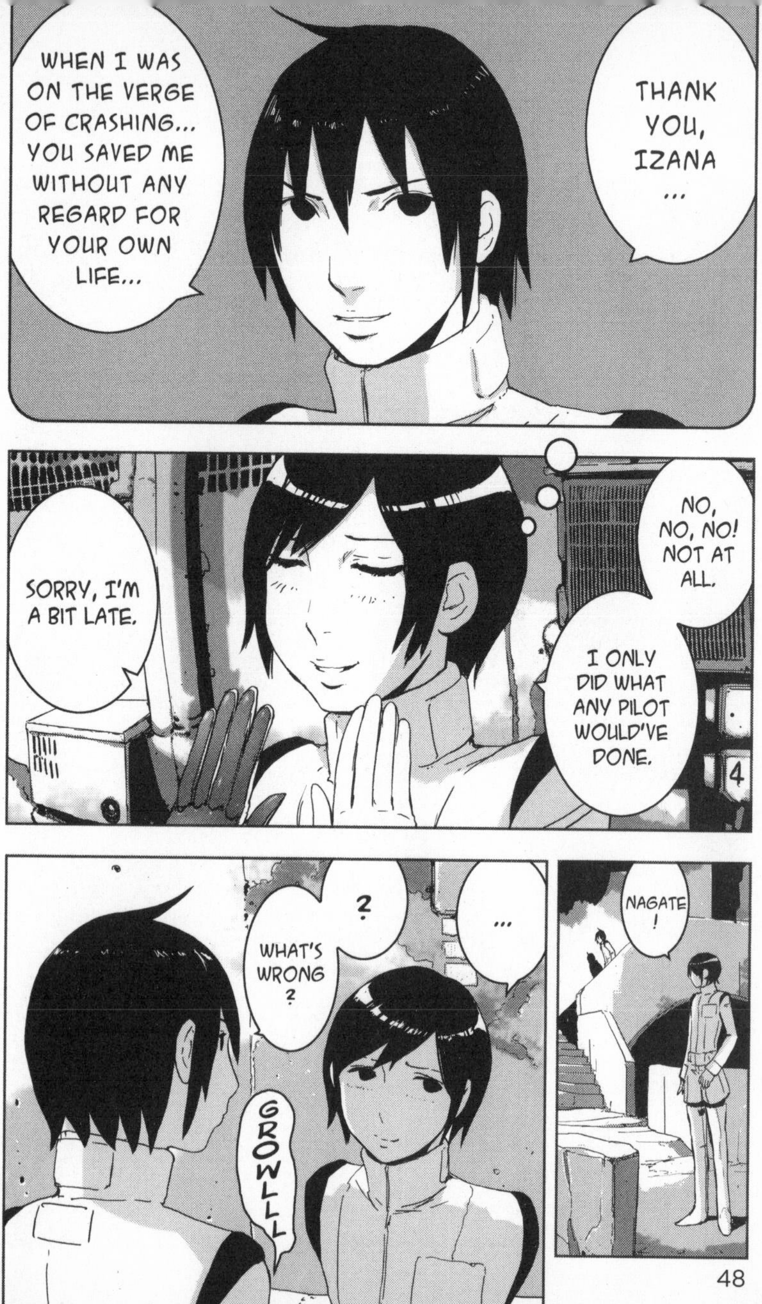
THANK YOU, IZANA ...
WHEN I WAS ON THE VERGE OF CRASHING... YOU SAVED ME WITHOUT ANY REGARD FOR YOUR OWN LIFE...
NO, NO, NO! NOT AT ALL.
I ONLY DID WHAT ANY PILOT WOULD'VE DONE.
4
SORRY, I'M A BIT LATE.
NAGATE !
...
?
WHAT'S WRONG ?
GROWLLL

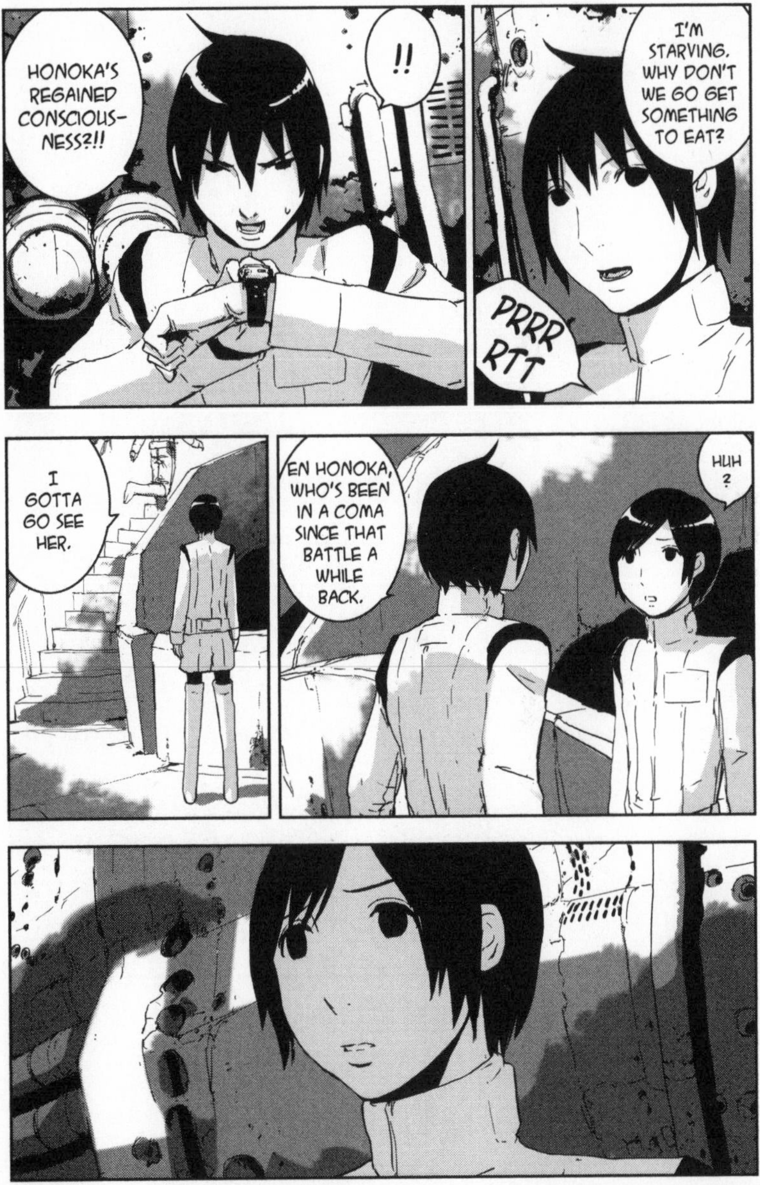
I'M STARVING. WHY DON'T WE GO GET SOMETHING TO EAT?
PRRRR RTT
!!
HONOKA'S REGAINED CONSCIOUS-NESS?!!
HUH?
EN HONOKA, WHO'S BEEN IN A COMA SINCE THAT BATTLE A WHILE BACK.
I GOTTA GO SEE HER.

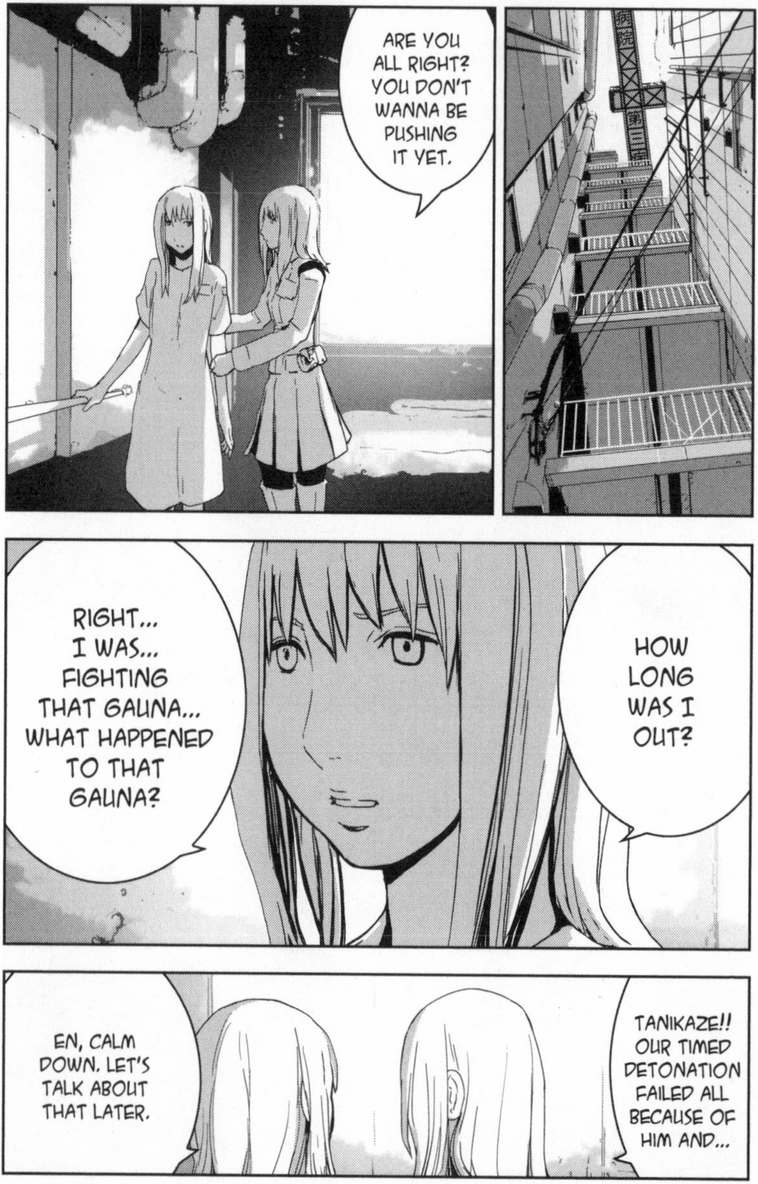
ARE YOU ALL RIGHT? YOU DON'T WANNA BE PUSHING IT YET.
HOW LONG WAS I OUT?
RIGHT... I WAS... FIGHTING THAT GAUNA... WHAT HAPPENED TO THAT GAUNA?
TANIKAZE!! OUR TIMED DETONATION FAILED ALL BECAUSE OF HIM AND...
EN, CALM DOWN. LET'S TALK ABOUT THAT LATER.

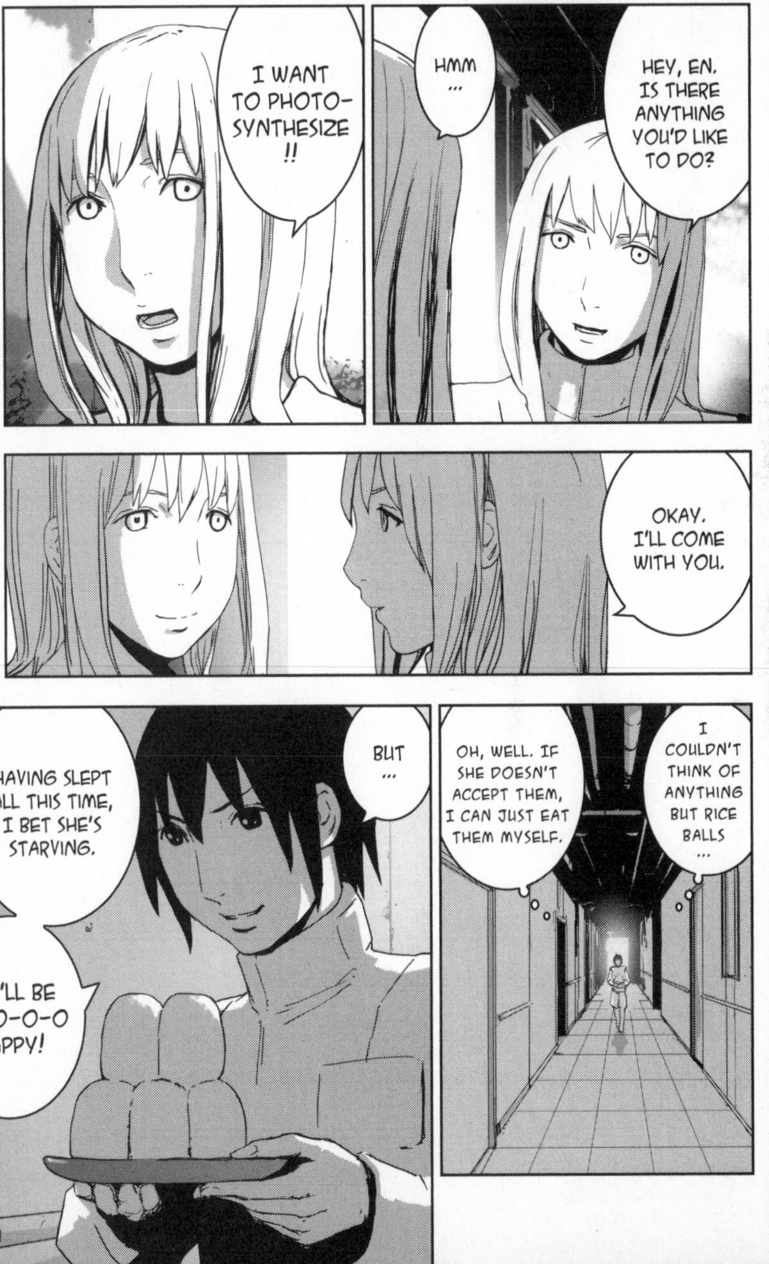
HEY, EN. IS THERE ANYTHING YOU'D LIKE TO DO?
HMM ...
I WANT TO PHOTO-SYNTHESIZE !!
OKAY. I'LL COME WITH YOU.
I COULDN'T THINK OF ANYTHING BUT RICE BALLS ...
OH, WELL. IF SHE DOESN'T ACCEPT THEM, I CAN JUST EAT THEM MYSELF.
BUT ...
HAVING SLEPT ALL THIS TIME, I BET SHE'S STARVING.
SHE'LL BE SO-O-O-O HAPPY!

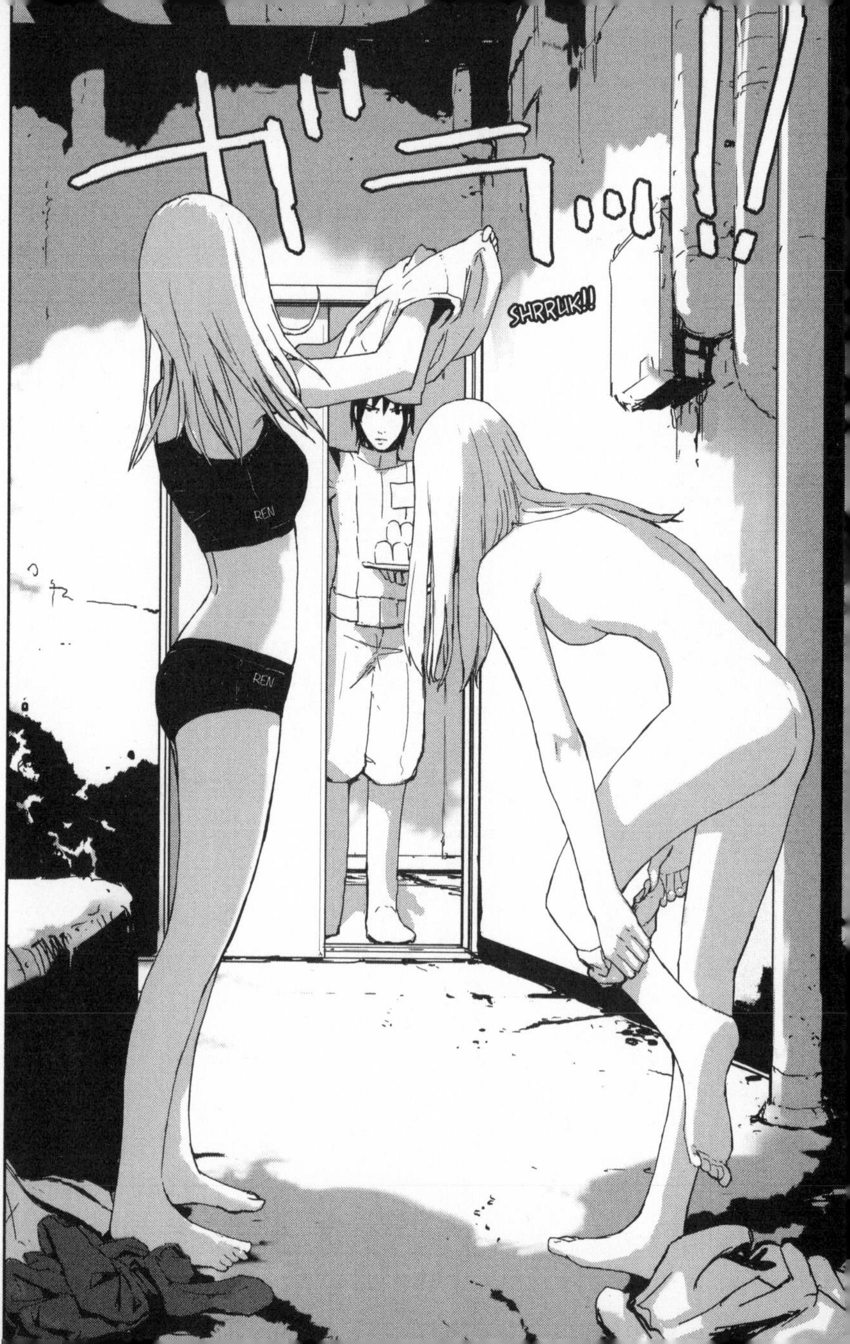
ガラ…!!
SHRRUK!!
REN
REN

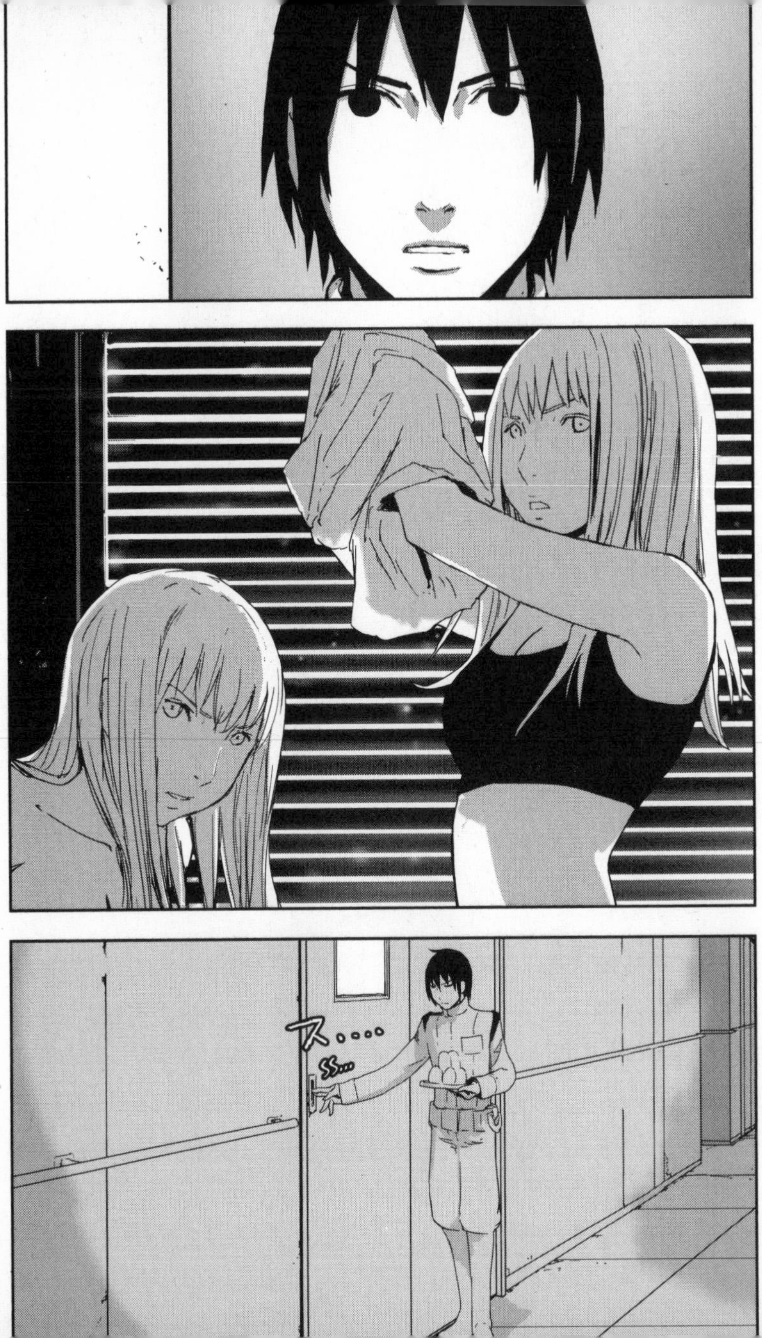
ス....
SS...

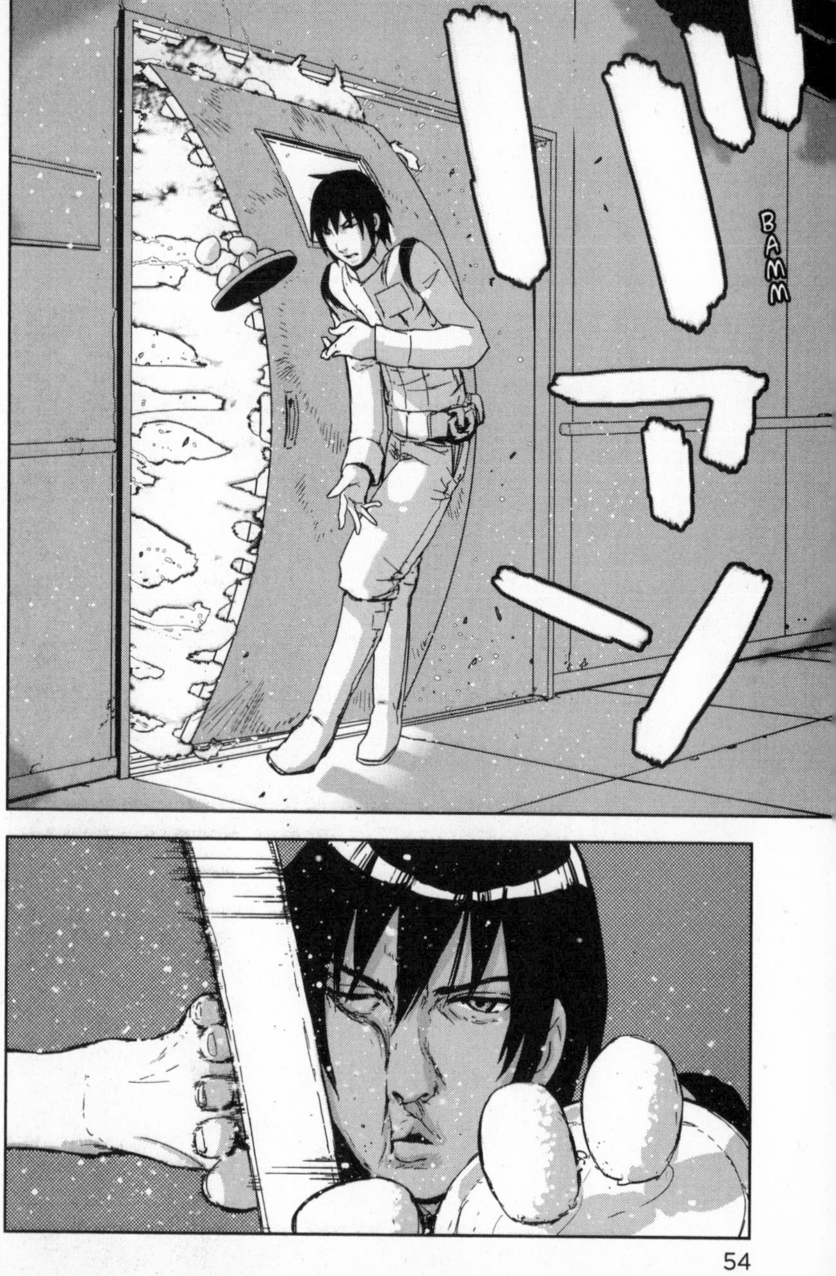
BAMM

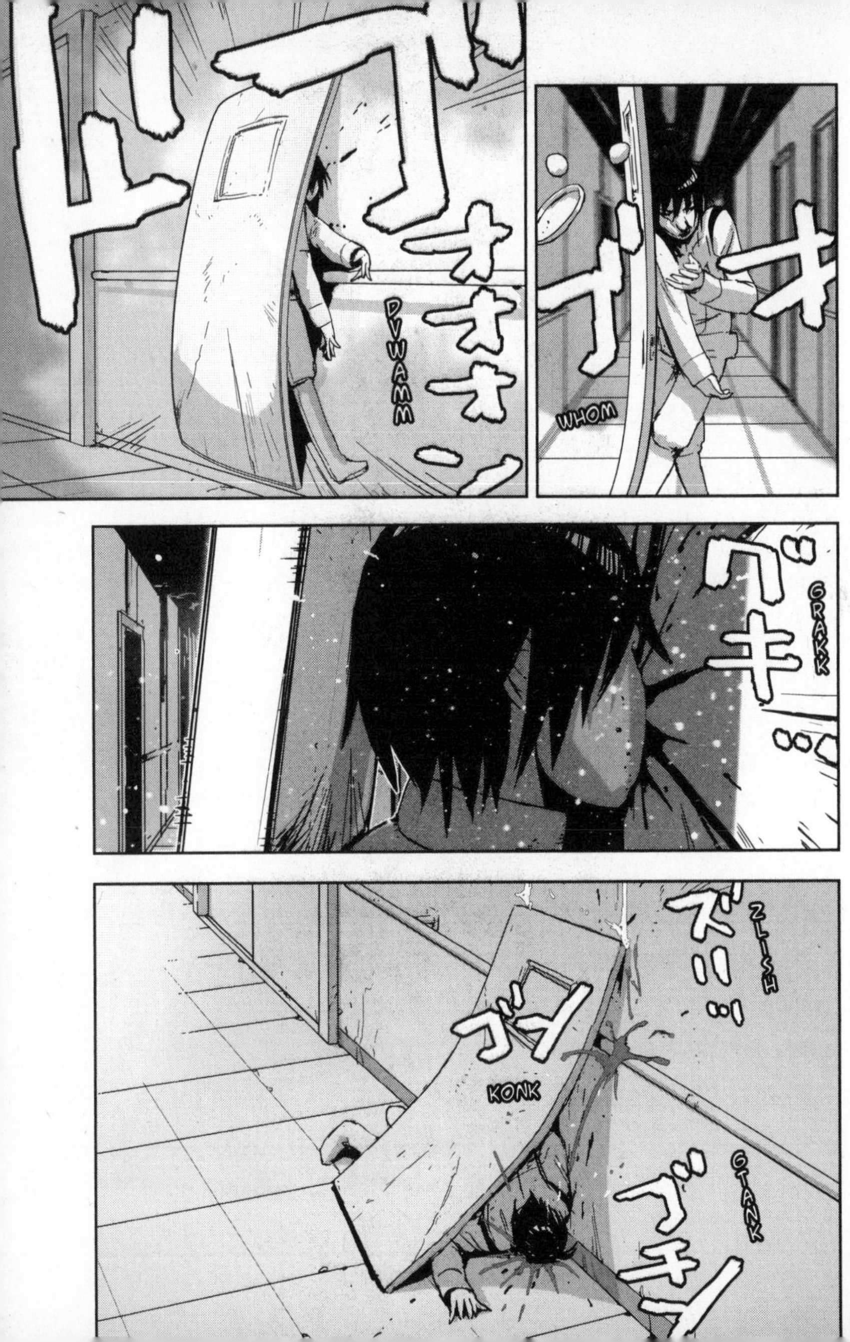
WHOM
DVWAMM
GRAKK
ZLISH
KONK
GTANK

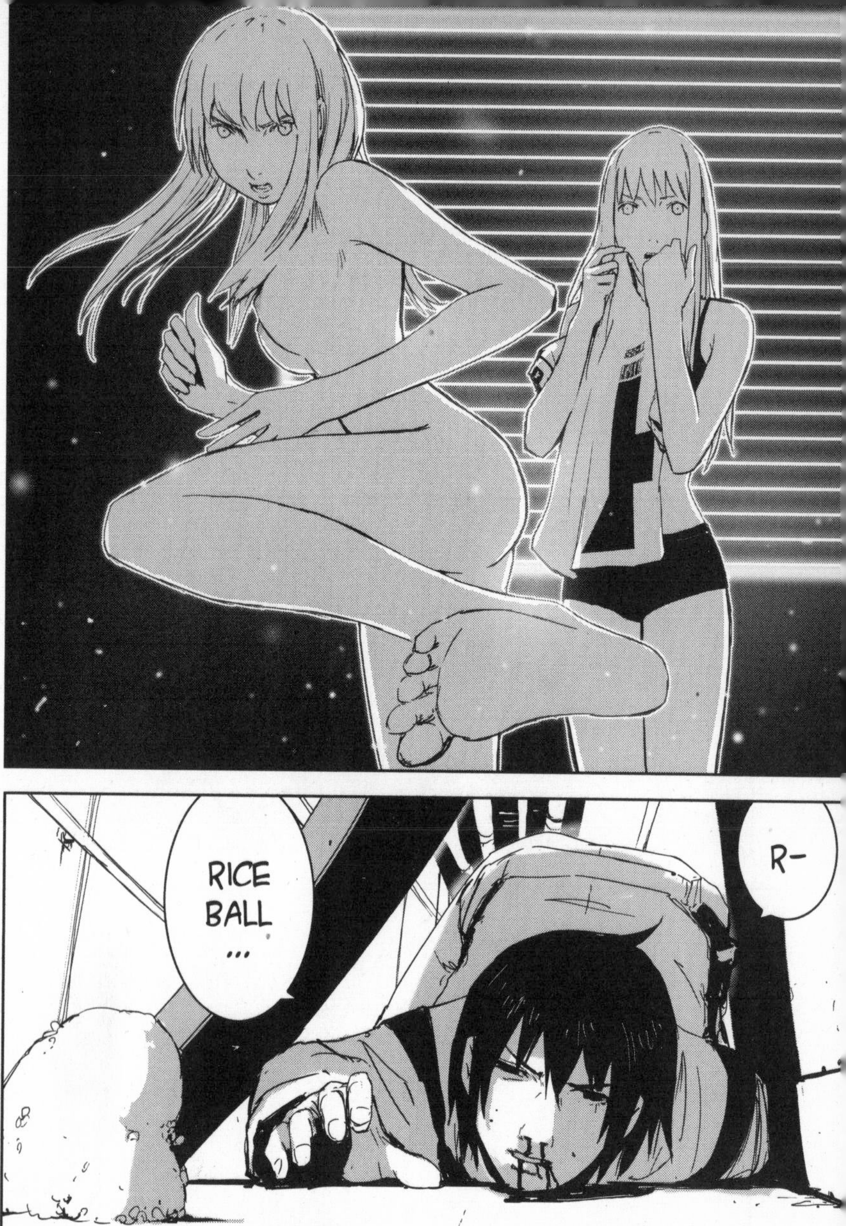
R-
RICE BALL ...

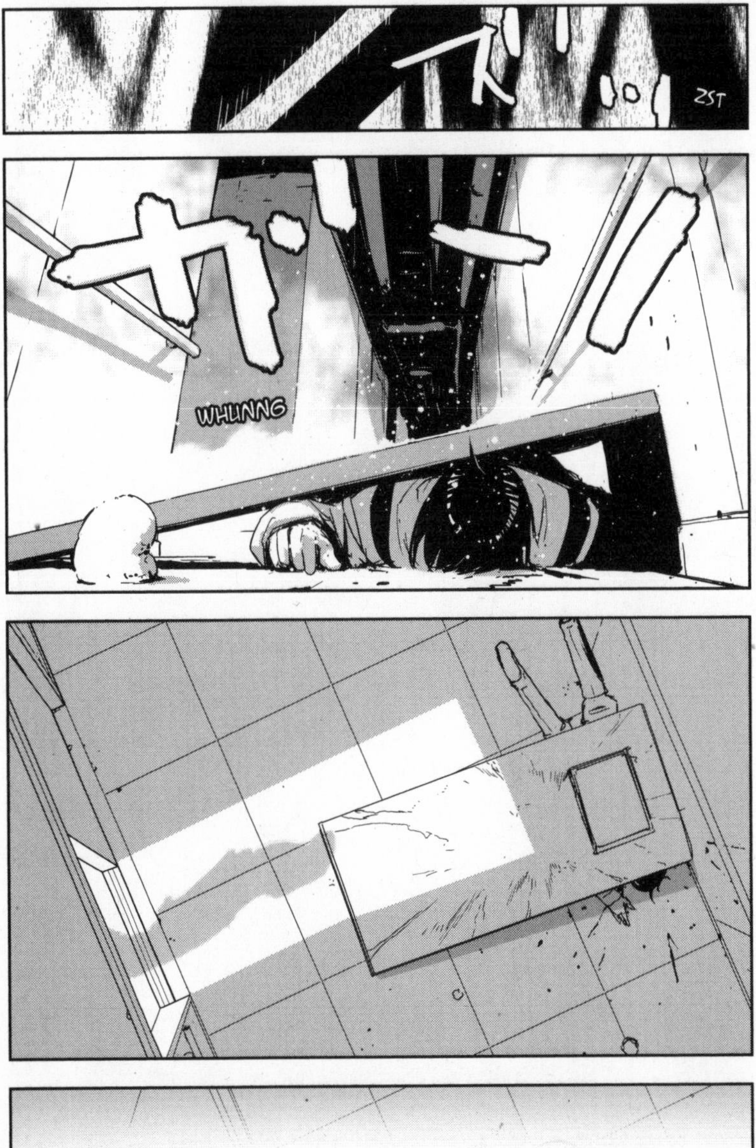
ZST
WHUNNG

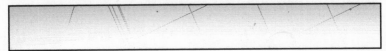

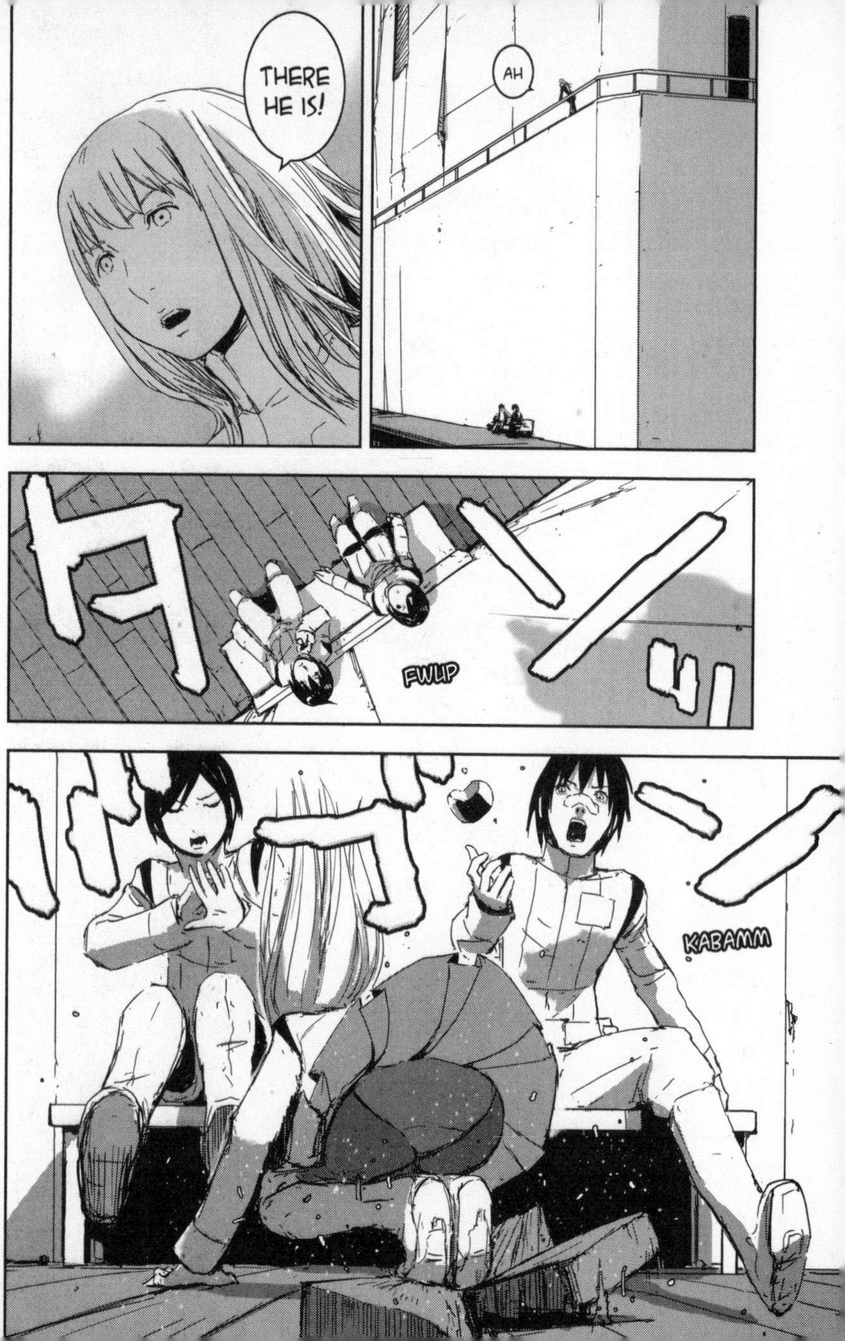
AH
THERE HE IS!
FWUP
KABAMM

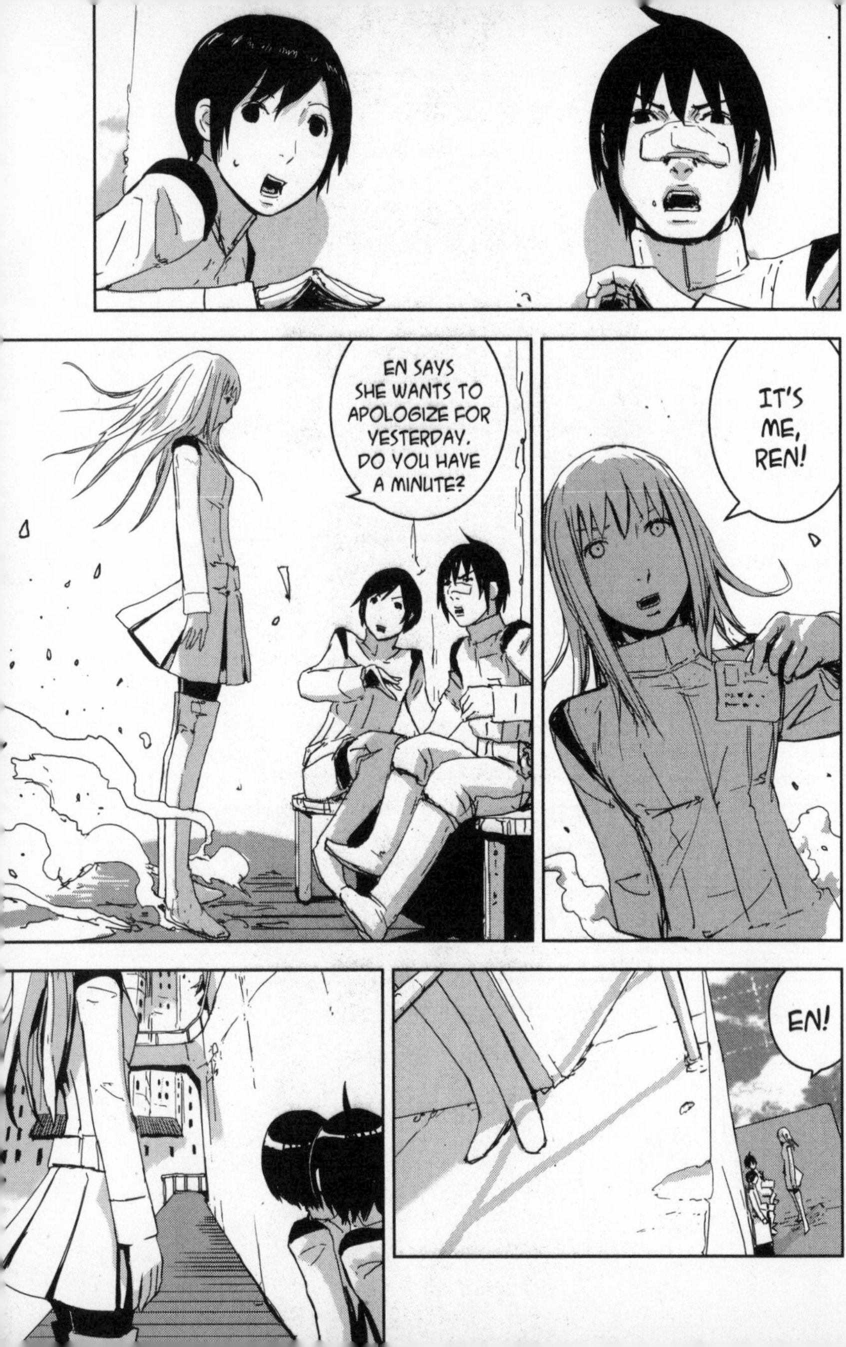
IT'S ME, REN!
EN SAYS SHE WANTS TO APOLOGIZE FOR YESTERDAY. DO YOU HAVE A MINUTE?
EN!

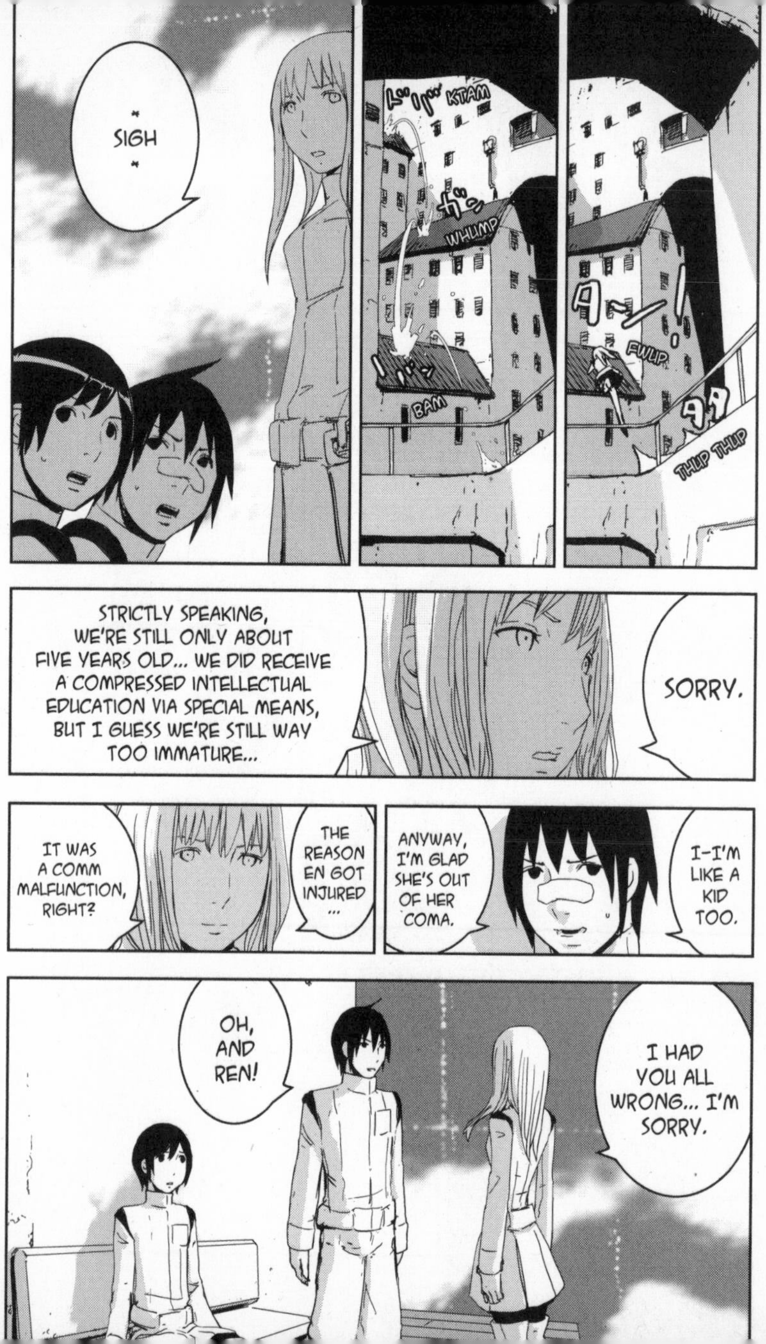
FWLP
THUP THUP
KTAM
WHUMP
BAM
SIGH
SORRY.
STRICTLY SPEAKING, WE'RE STILL ONLY ABOUT FIVE YEARS OLD... WE DID RECEIVE A COMPRESSED INTELLECTUAL EDUCATION VIA SPECIAL MEANS, BUT I GUESS WE'RE STILL WAY TOO IMMATURE...
I-I'M LIKE A KID TOO.
ANYWAY, I'M GLAD SHE'S OUT OF HER COMA.
THE REASON EN GOT INJURED ...
IT WAS A COMM MALFUNCTION, RIGHT?
I HAD YOU ALL WRONG... I'M SORRY.
OH, AND REN!

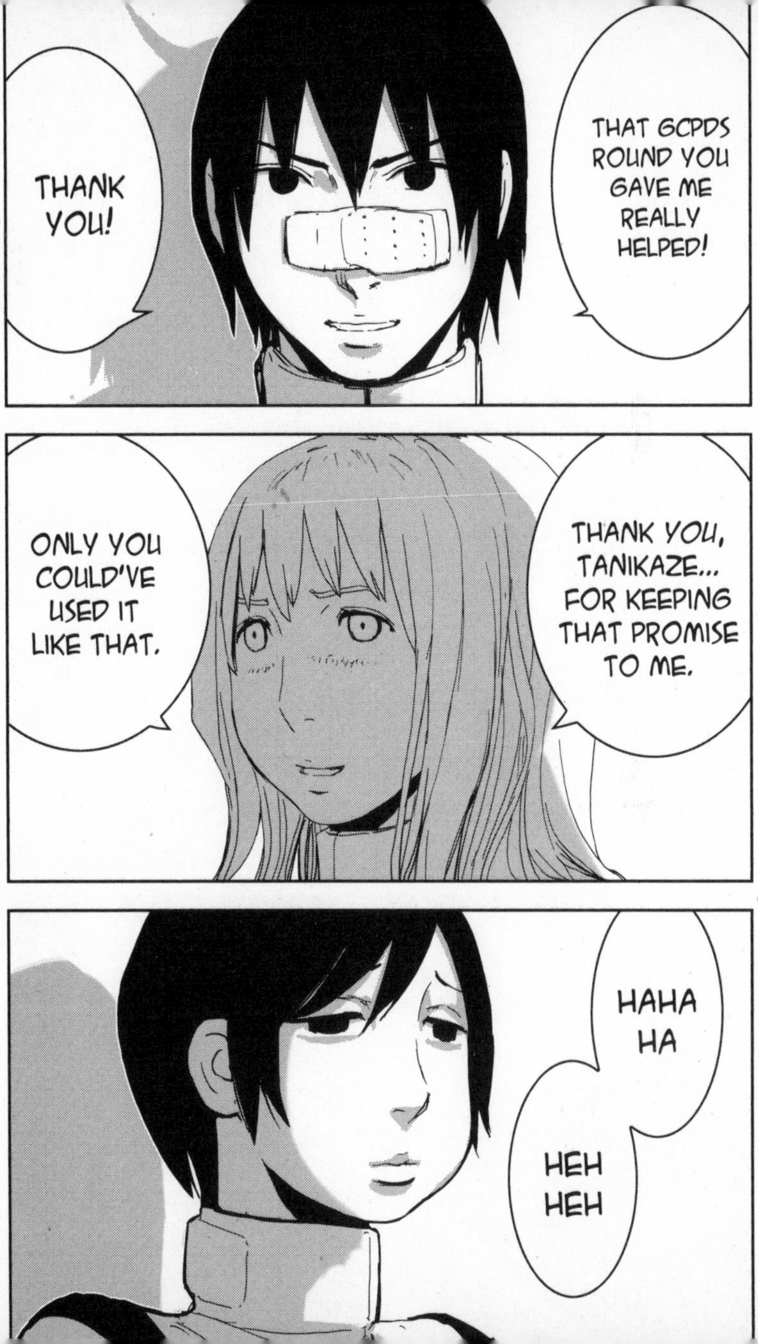
THAT GCPDS ROUND YOU GAVE ME REALLY HELPED!
THANK YOU!
THANK YOU, TANIKAZE... FOR KEEPING THAT PROMISE TO ME.
ONLY YOU COULD'VE USED IT LIKE THAT.
HAHA HA
HEH HEH

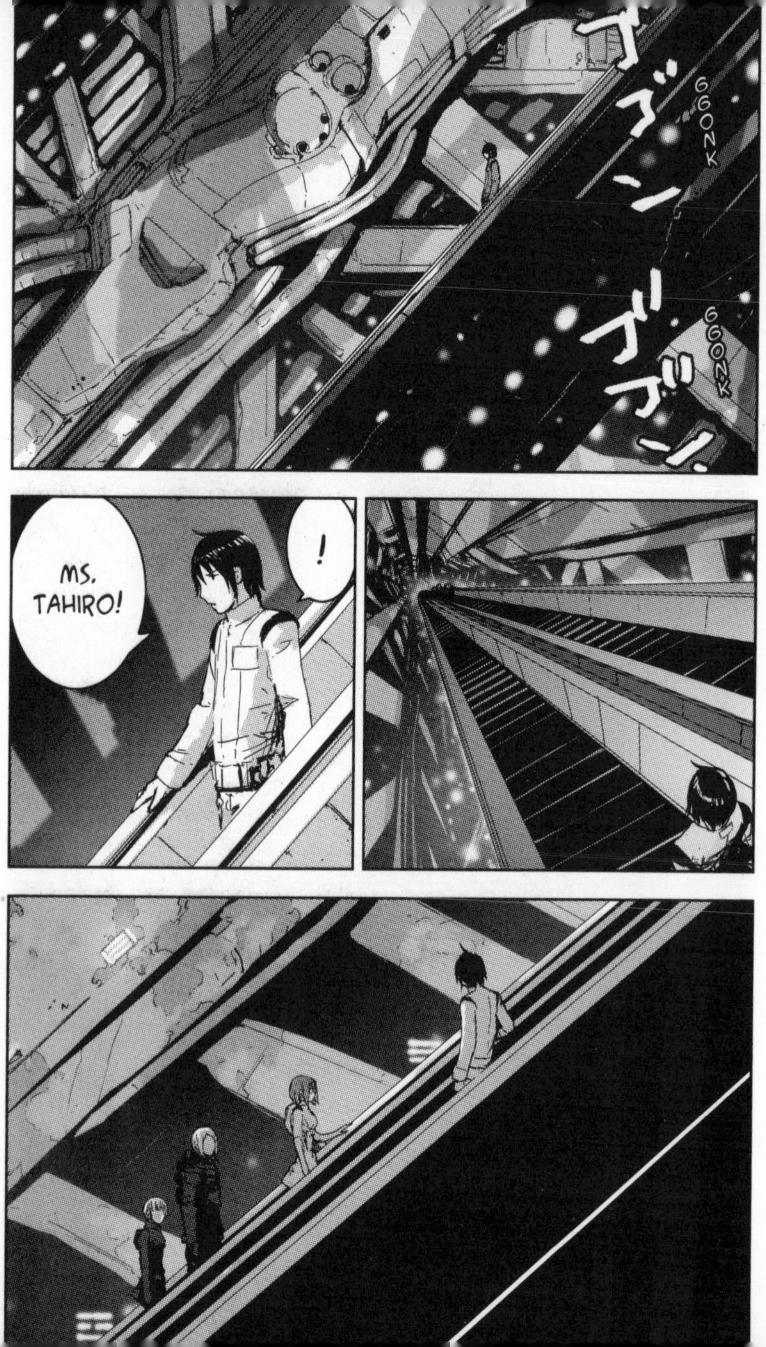
GGONK
GGONK
!
MS. TAHIRO!

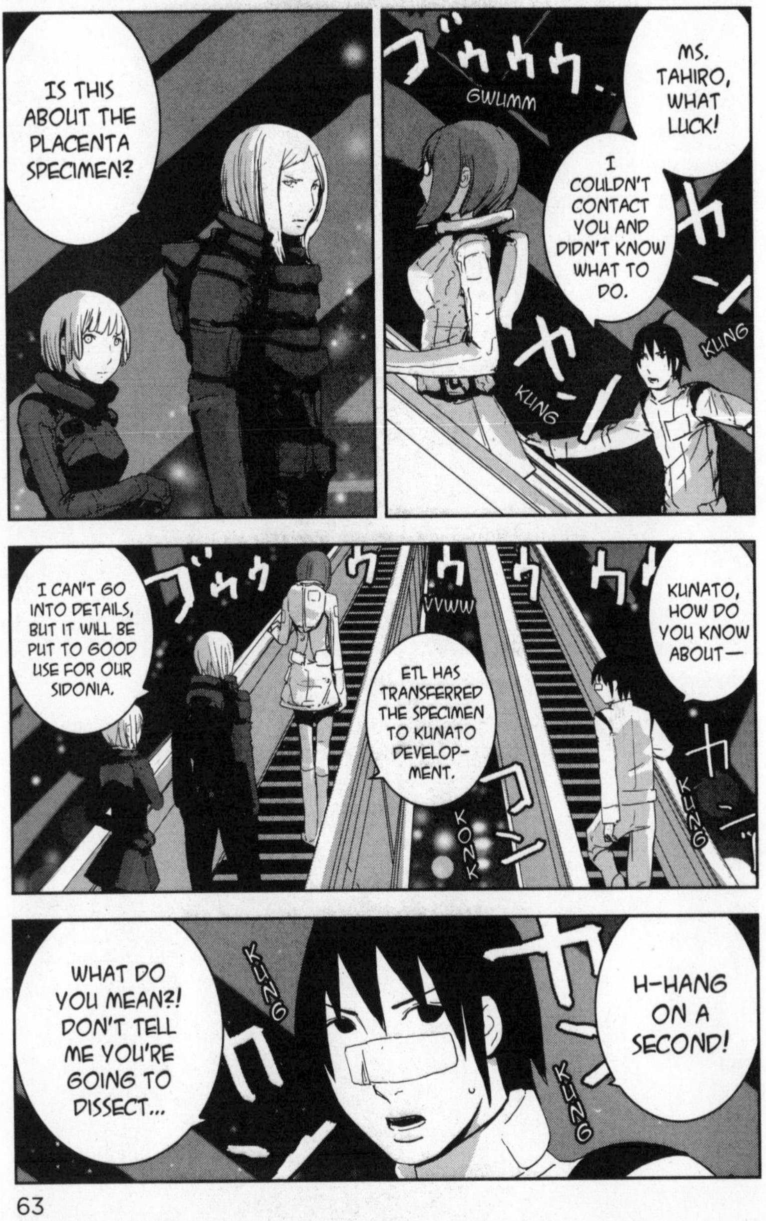
GWUMM
MS. TAHIRO, WHAT LUCK!
I COULDN'T CONTACT YOU AND DIDN'T KNOW WHAT TO DO.
KUNG
KUNG
IS THIS ABOUT THE PLACENTA SPECIMEN?
KUNATO, HOW DO YOU KNOW ABOUT—
VVWW
ETL HAS TRANSFERRED THE SPECIMEN TO KUNATO DEVELOPMENT.
KUNG
KONK
I CAN'T GO INTO DETAILS, BUT IT WILL BE PUT TO GOOD USE FOR OUR SIDONIA.
H-HANG ON A SECOND!
KUNG
KUNG
WHAT DO YOU MEAN?! DON'T TELL ME YOU'RE GOING TO DISSECT...

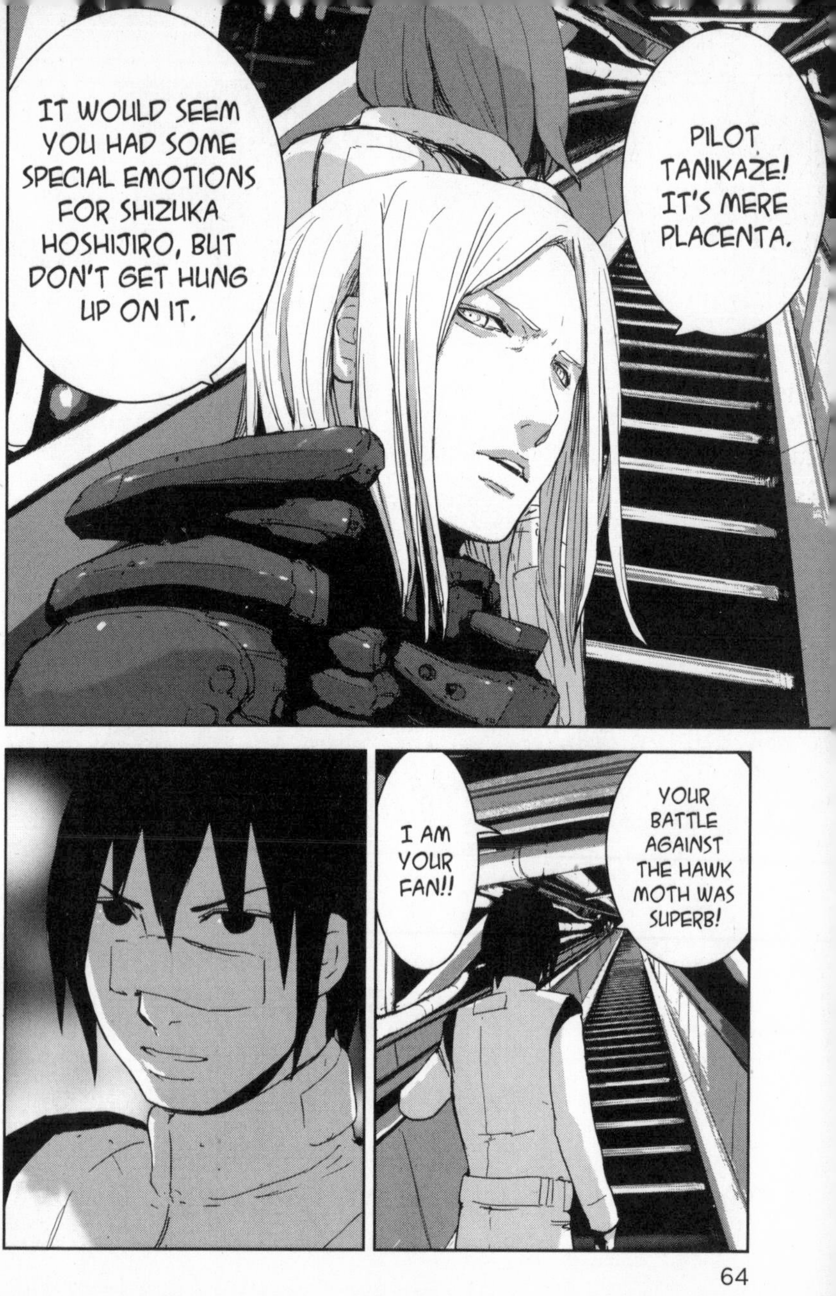
PILOT TANIKAZE! IT'S MERE PLACENTA.
IT WOULD SEEM YOU HAD SOME SPECIAL EMOTIONS FOR SHIZUKA HOSHIJIRO, BUT DON'T GET HUNG UP ON IT.
YOUR BATTLE AGAINST THE HAWK MOTH WAS SUPERB!
I AM YOUR FAN!!

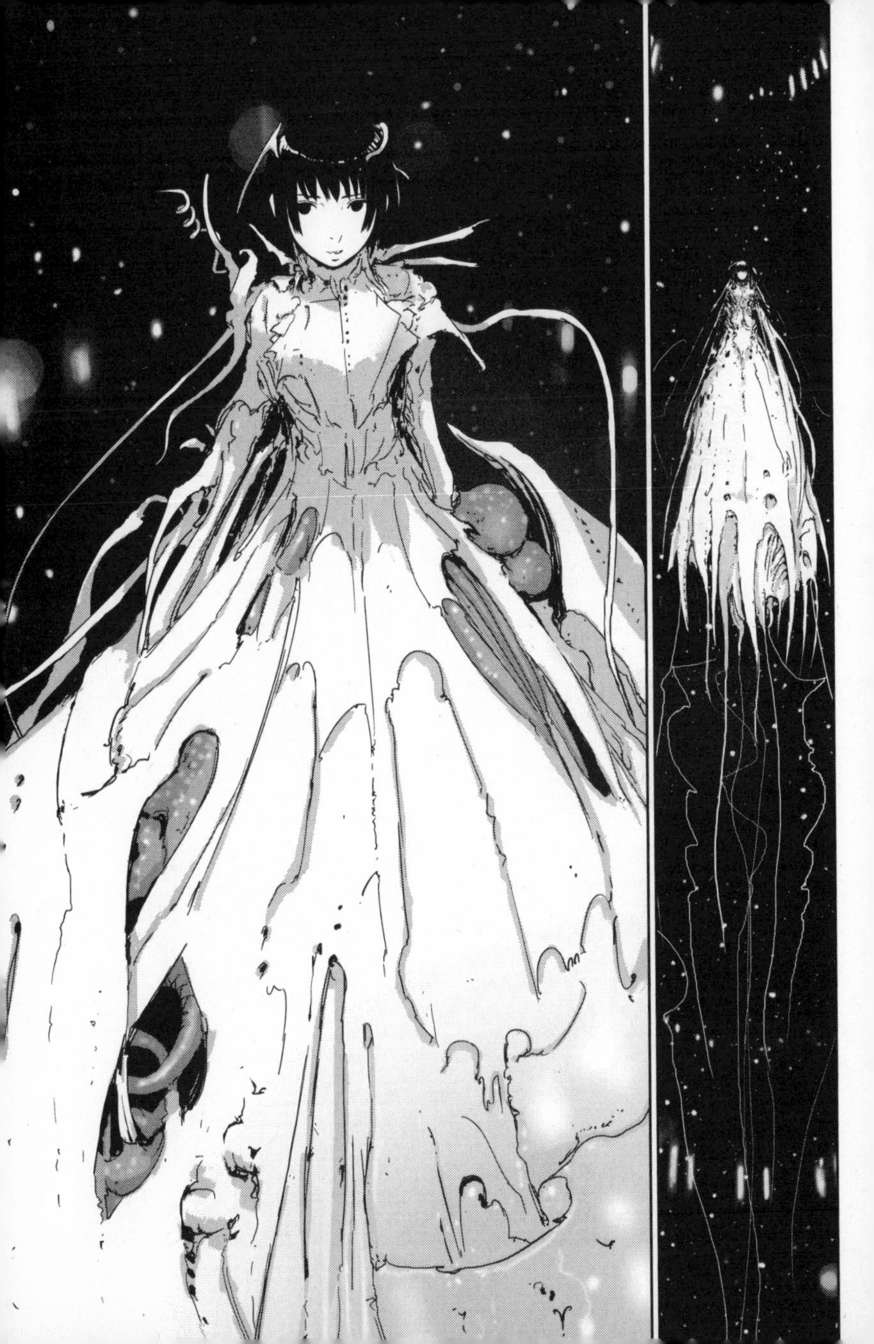

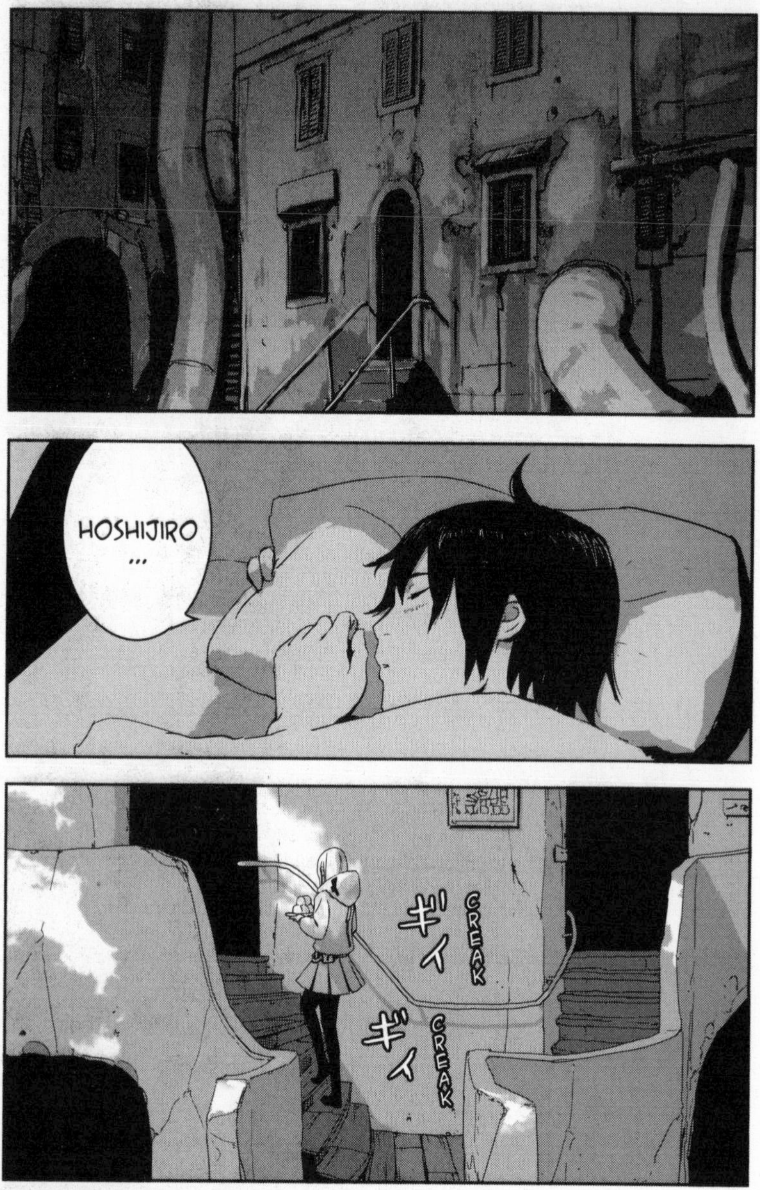
HOSHIJIRO ...
CREAK
ギイ
CREAK
ギイ

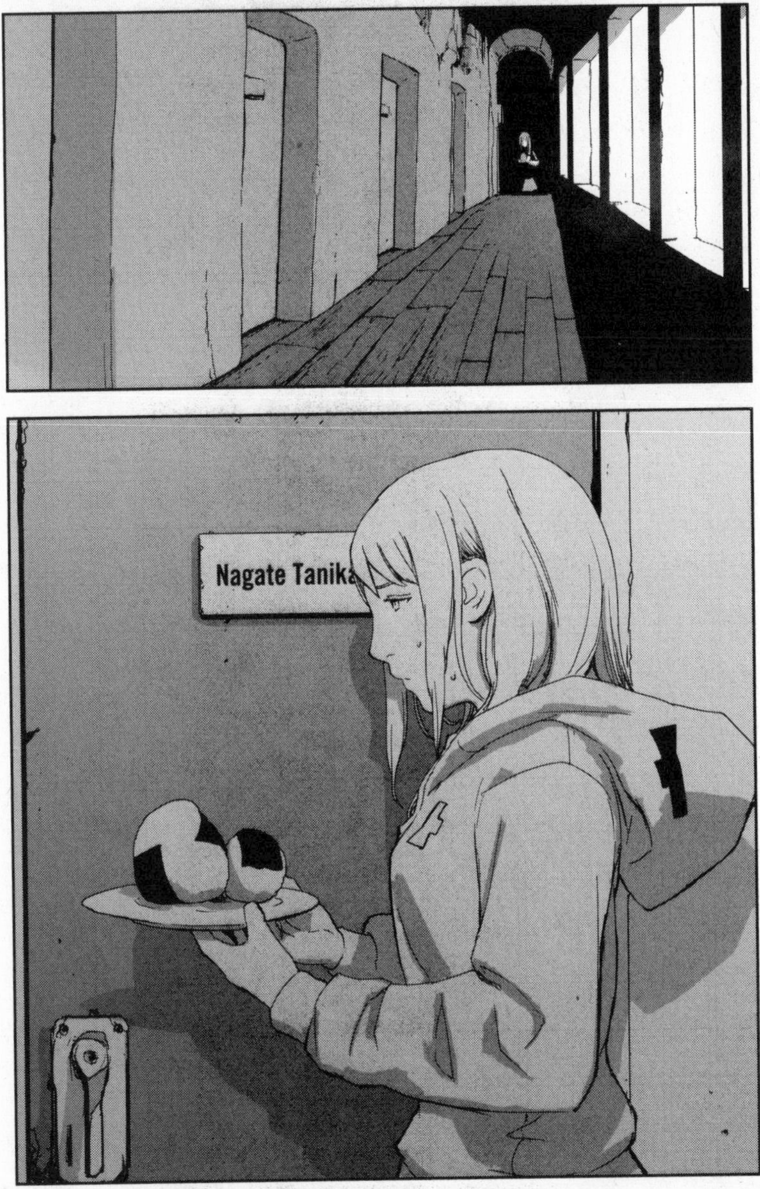
Nagate Tanika

Chapter 27: END

Chapter 28: Placenta-Hoshijiro's Impregnation

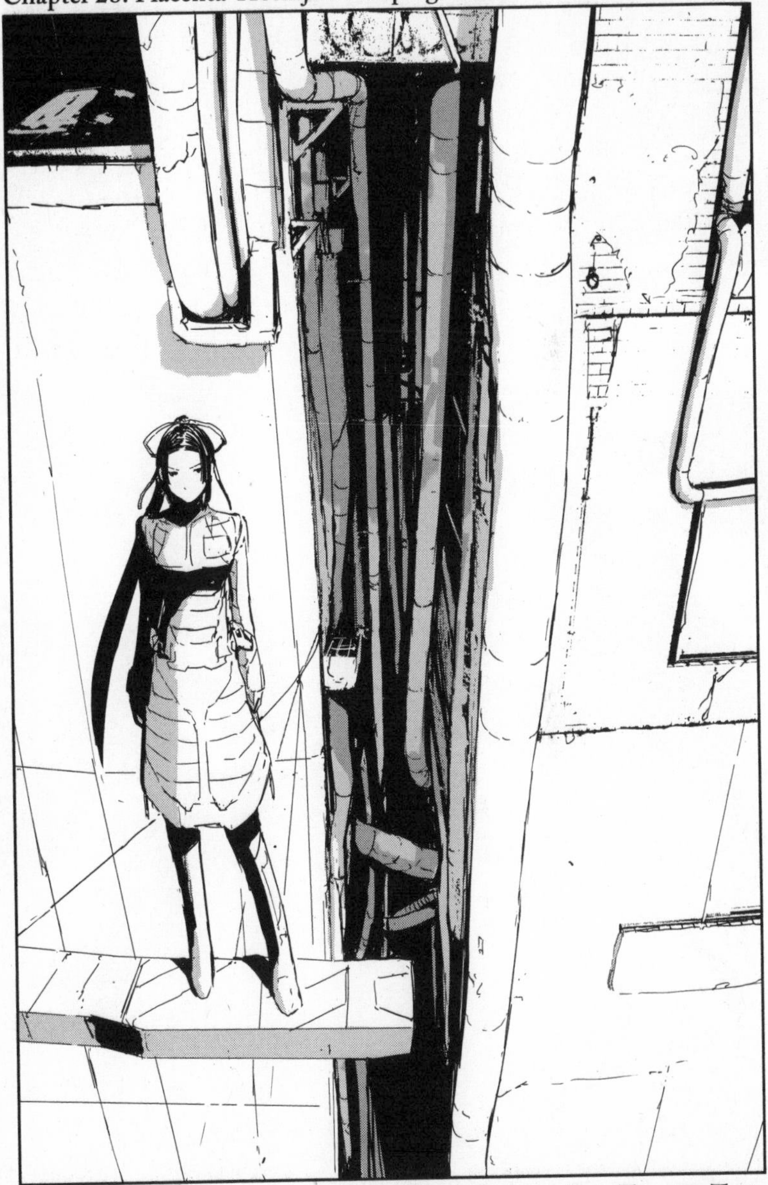

One Hundred Sights of Sidonia Part Twenty-Two: Residential Tower Top Level Outer Wall

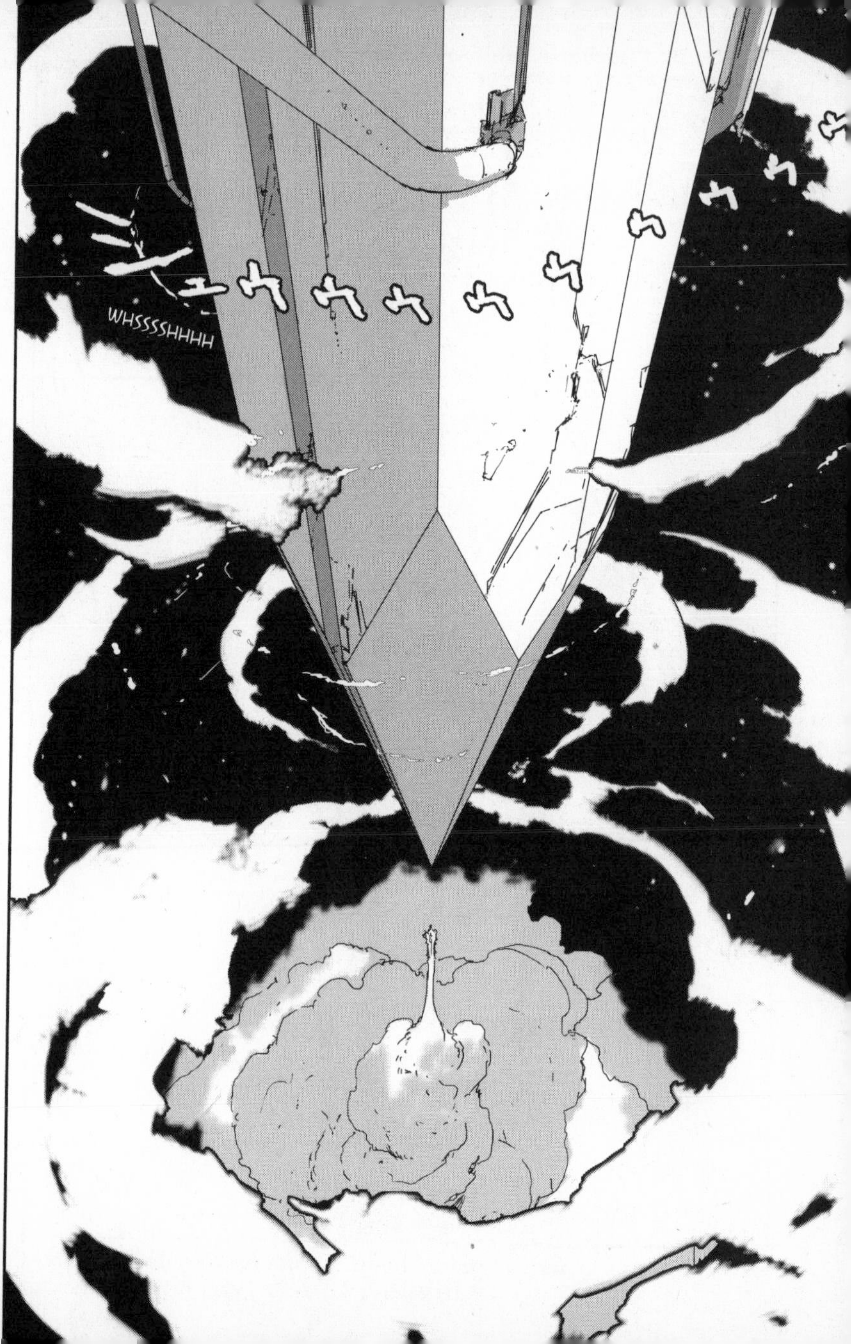
WHSSSSHHHH

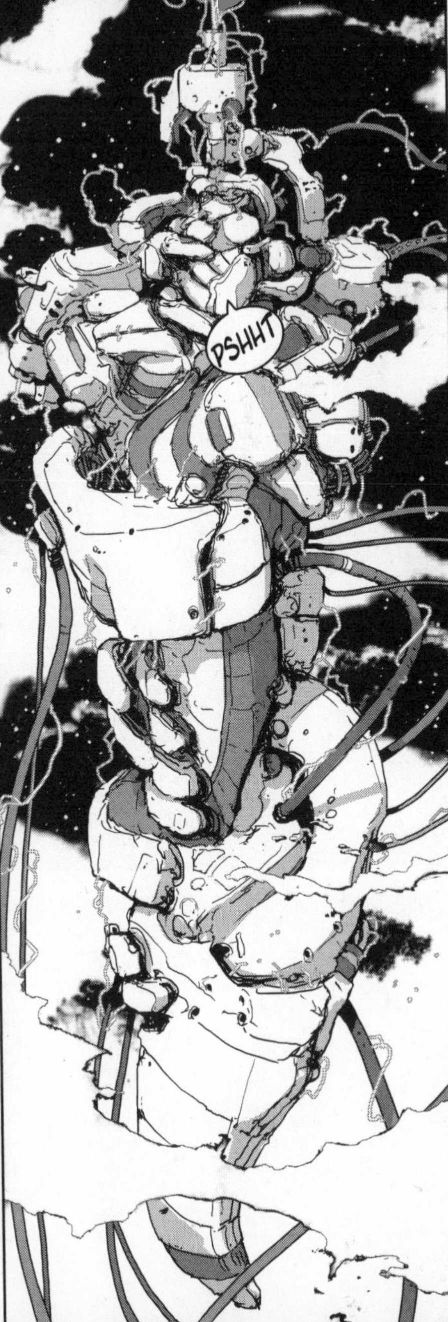
PSHHT
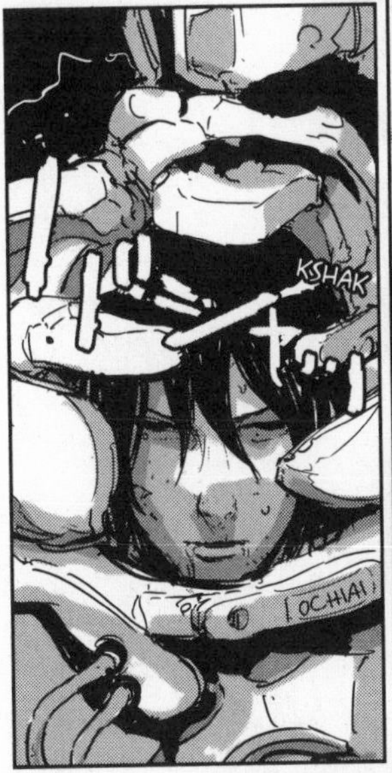
KSHAK
OCHIAI
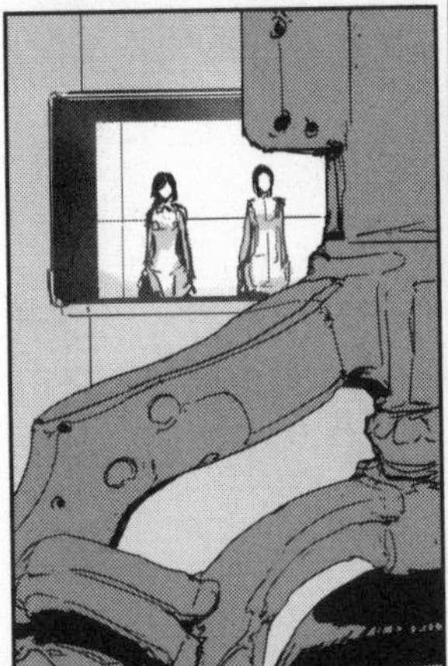

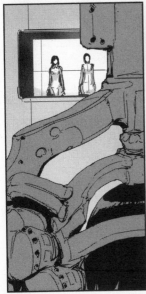

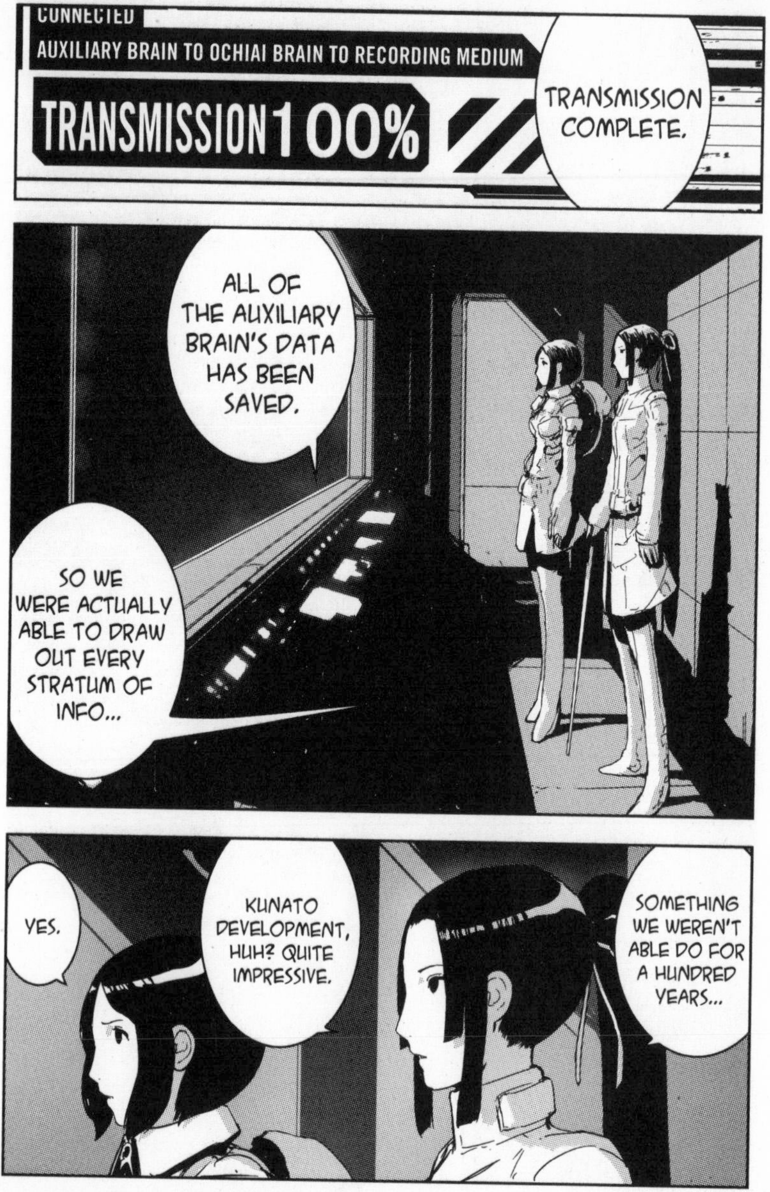

CONNECTED
AUXILIARY BRAIN TO OCHIAI BRAIN TO RECORDING MEDIUM
TRANSMISSION 100%
TRANSMISSION COMPLETE.
ALL OF THE AUXILIARY BRAIN'S DATA HAS BEEN SAVED.
SO WE WERE ACTUALLY ABLE TO DRAW OUT EVERY STRATUM OF INFO...
SOMETHING WE WEREN'T ABLE DO FOR A HUNDRED YEARS...
KUNATO DEVELOPMENT, HUH? QUITE IMPRESSIVE.
YES.

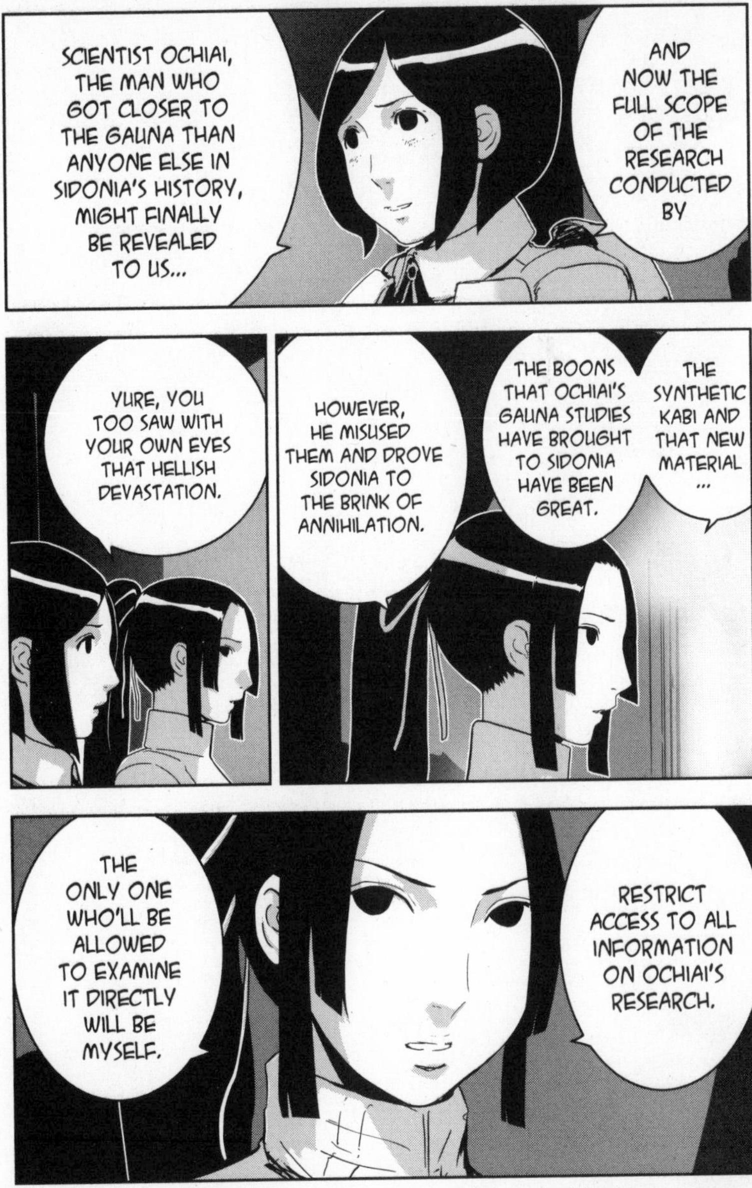
AND NOW THE FULL SCOPE OF THE RESEARCH CONDUCTED BY
SCIENTIST OCHIAI, THE MAN WHO GOT CLOSER TO THE GAUNA THAN ANYONE ELSE IN SIDONIA'S HISTORY, MIGHT FINALLY BE REVEALED TO US...
THE SYNTHETIC KABI AND THAT NEW MATERIAL ...
THE BOONS THAT OCHIAI'S GAUNA STUDIES HAVE BROUGHT TO SIDONIA HAVE BEEN GREAT.
HOWEVER, HE MISUSED THEM AND DROVE SIDONIA TO THE BRINK OF ANNIHILATION.
YURE, YOU TOO SAW WITH YOUR OWN EYES THAT HELLISH DEVASTATION.
RESTRICT ACCESS TO ALL INFORMATION ON OCHIAI'S RESEARCH.
THE ONLY ONE WHO'LL BE ALLOWED TO EXAMINE IT DIRECTLY WILL BE MYSELF.

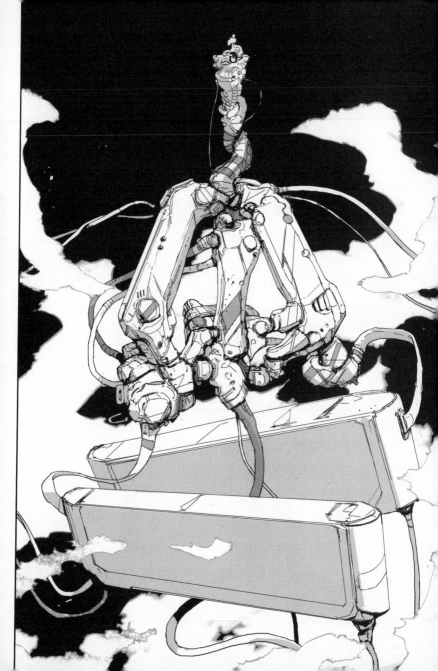

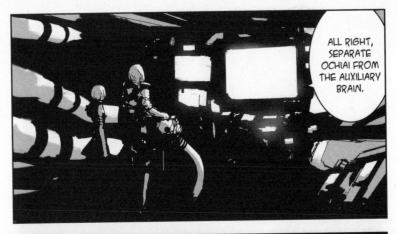

ALL RIGHT, SEPARATE OCHIAI FROM THE AUXILIARY BRAIN.

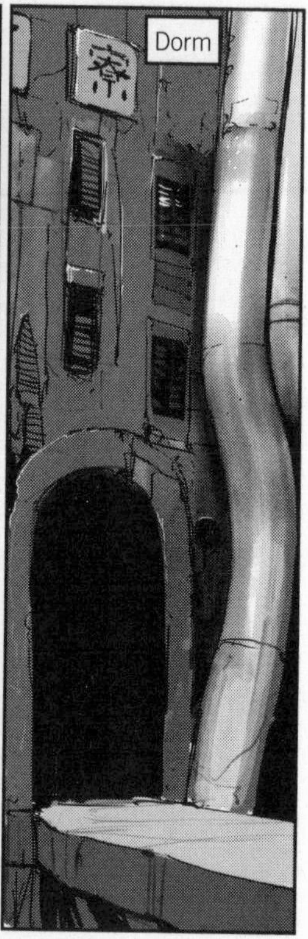
寮
Dorm

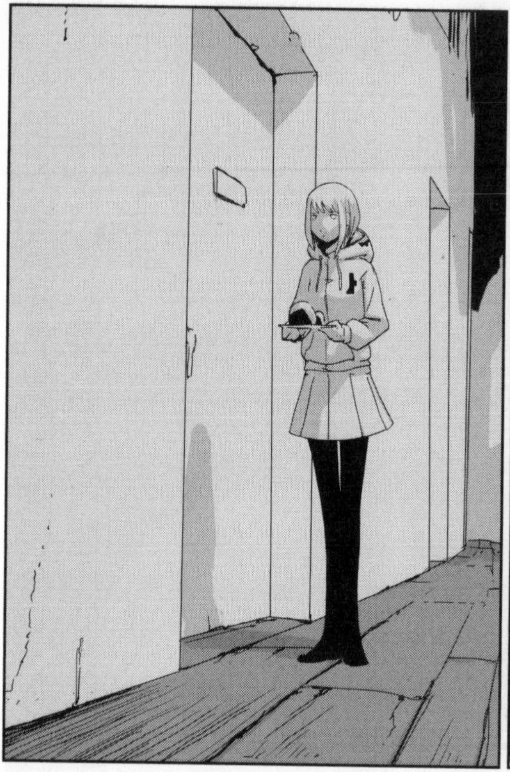

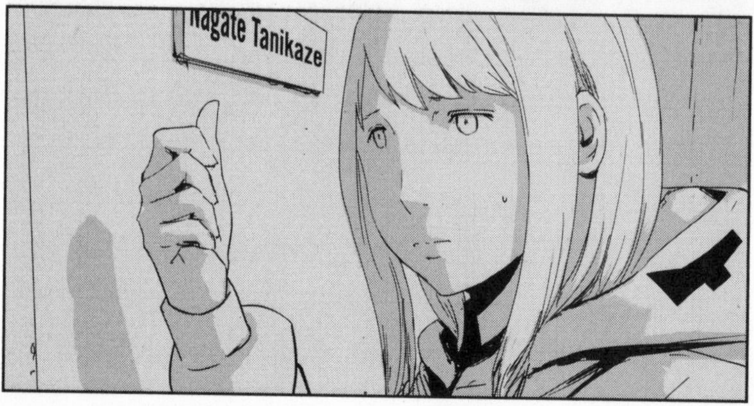
Nagate Tanikaze

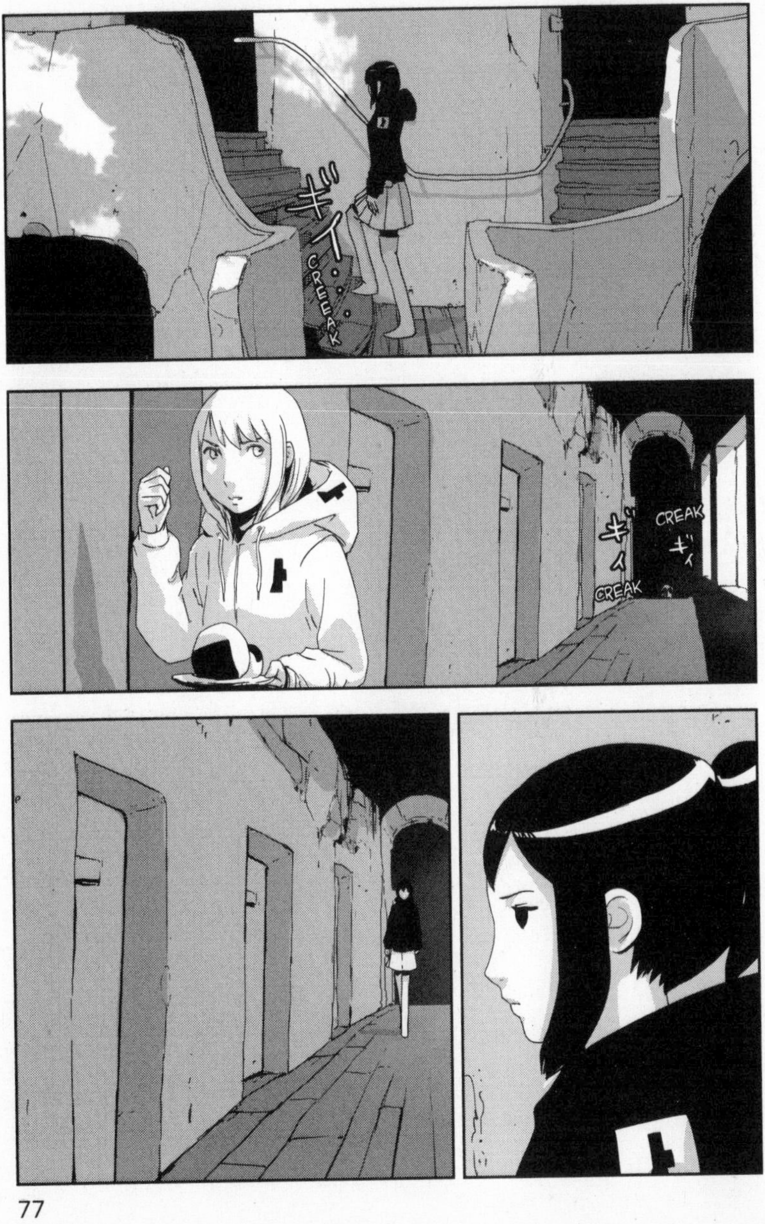
ギイ
CREEAK
CREAK
ギィ
ギイ
CREAK

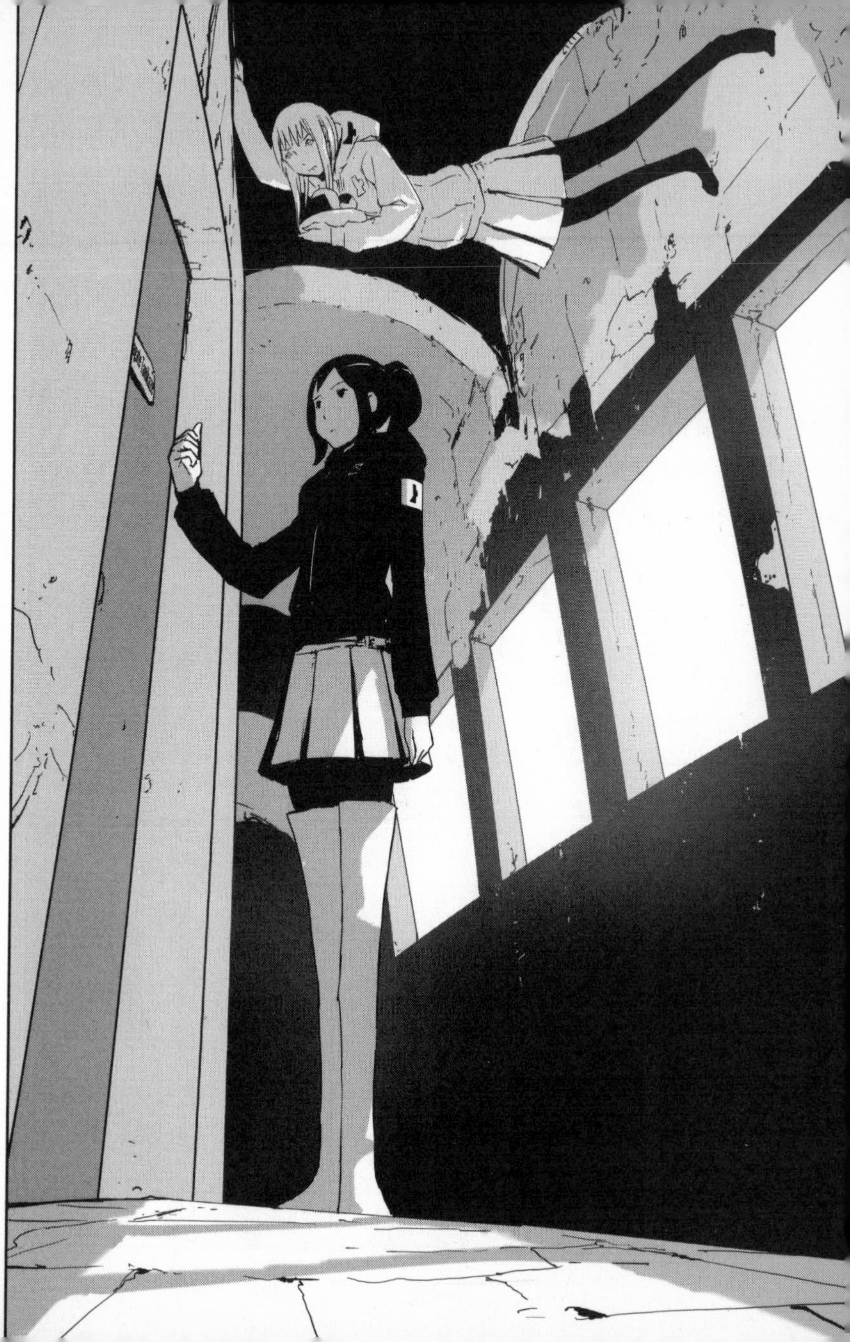

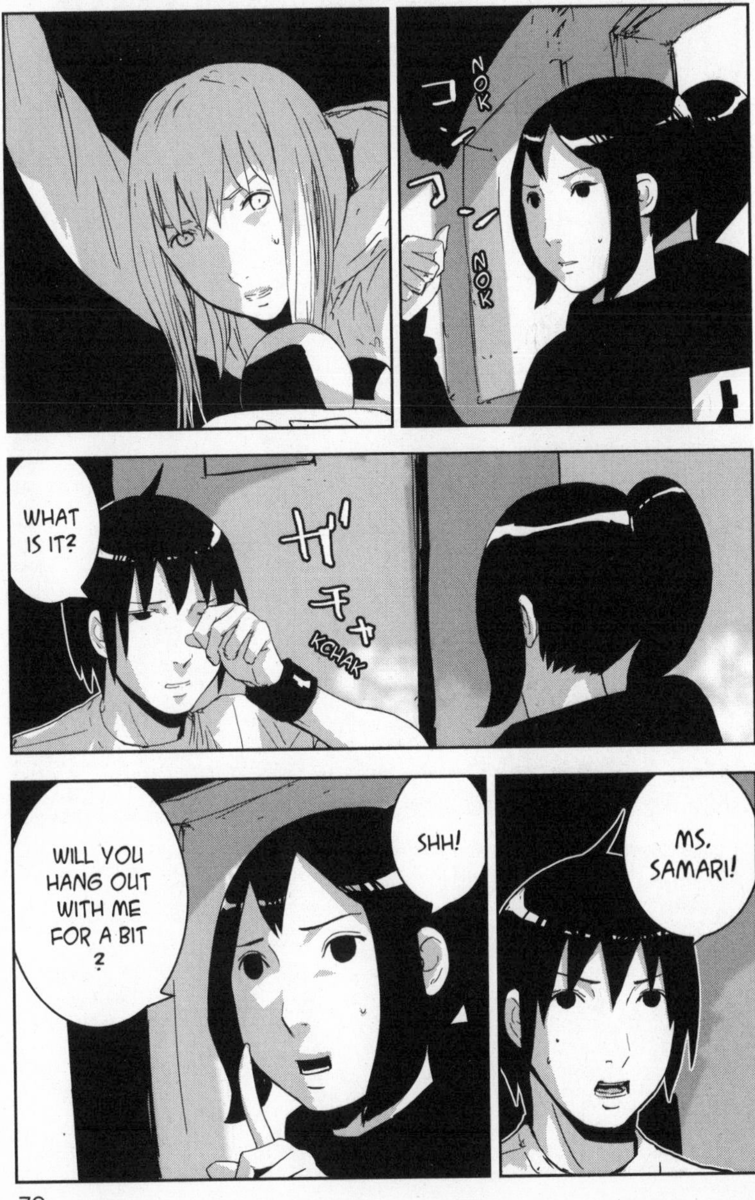
NOK
NOK
KCHAK
WHAT IS IT?
MS. SAMARI!
SHH!
WILL YOU HANG OUT WITH ME FOR A BIT ?

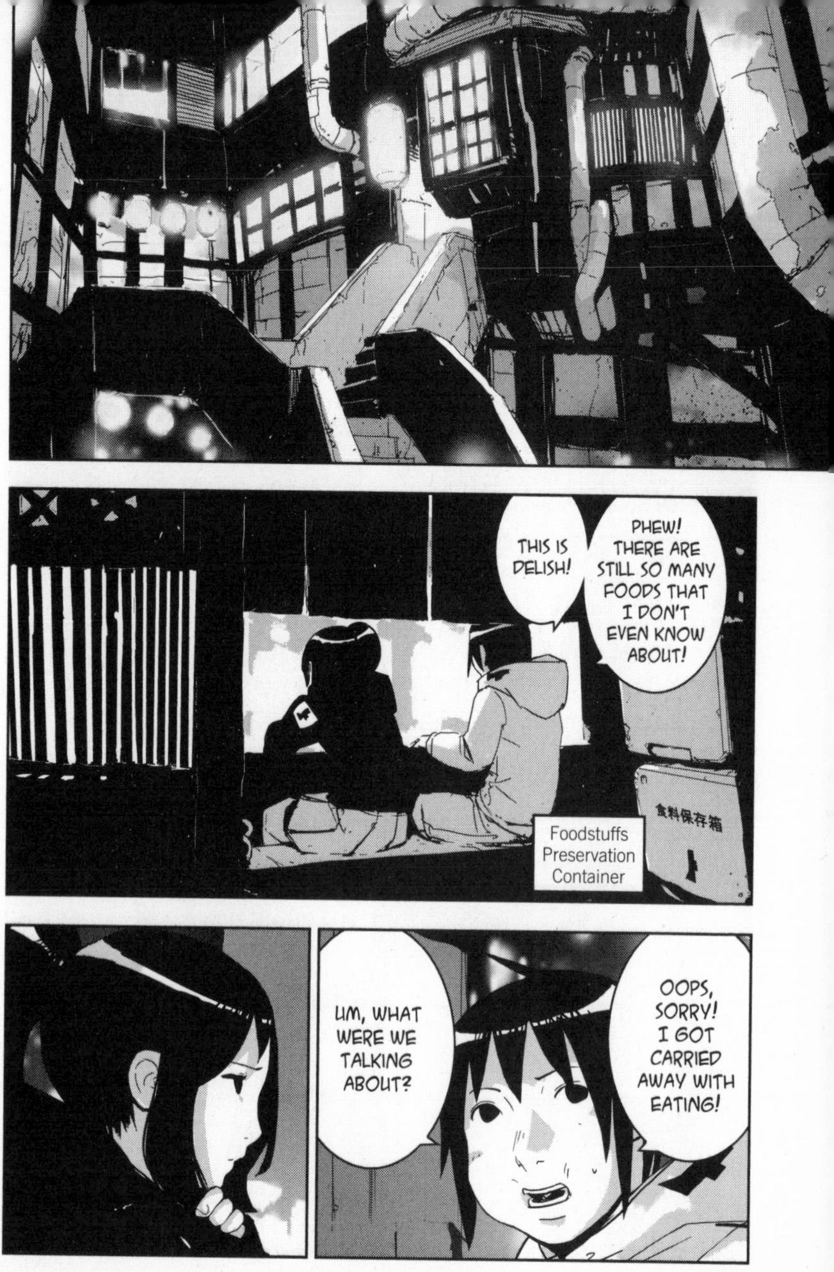
PHEW! THERE ARE STILL SO MANY FOODS THAT I DON'T EVEN KNOW ABOUT!
THIS IS DELISH!
食料保存箱
Foodstuffs Preservation Container
OOPS, SORRY! I GOT CARRIED AWAY WITH EATING!
UM, WHAT WERE WE TALKING ABOUT?

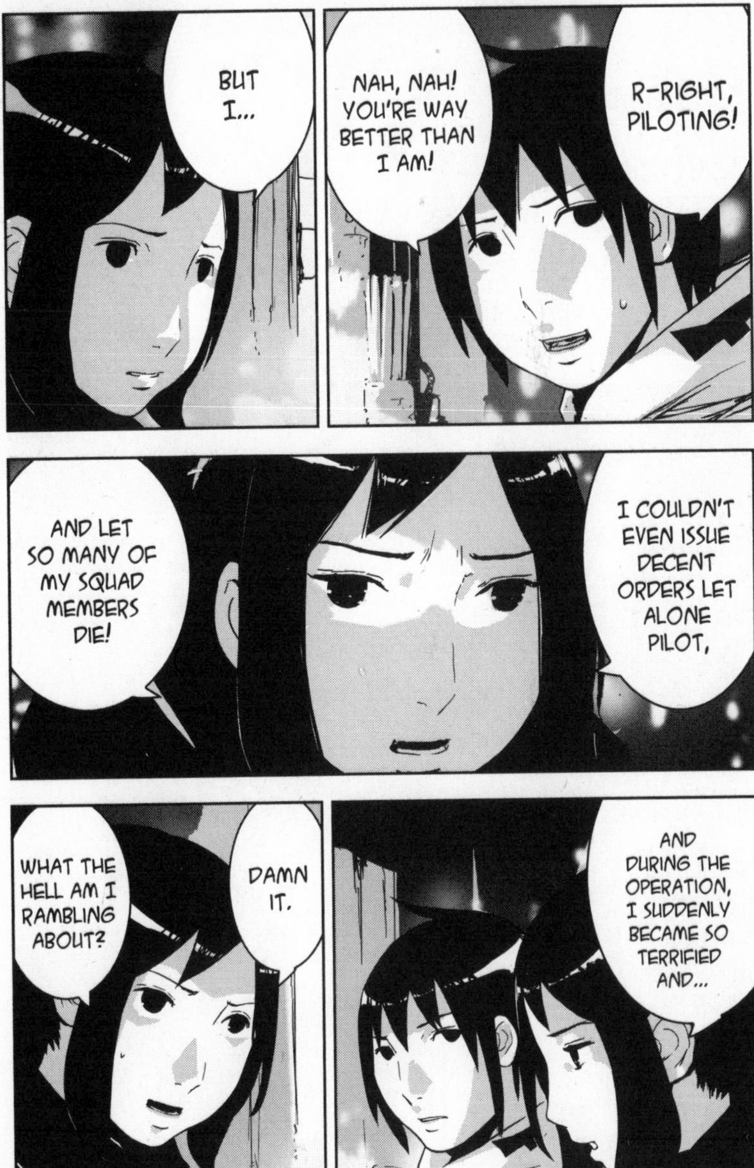
R-RIGHT, PILOTING!
NAH, NAH! YOU'RE WAY BETTER THAN I AM!
BUT I...
I COULDN'T EVEN ISSUE DECENT ORDERS LET ALONE PILOT,
AND LET SO MANY OF MY SQUAD MEMBERS DIE!
AND DURING THE OPERATION, I SUDDENLY BECAME SO TERRIFIED AND...
DAMN IT.
WHAT THE HELL AM I RAMBLING ABOUT?

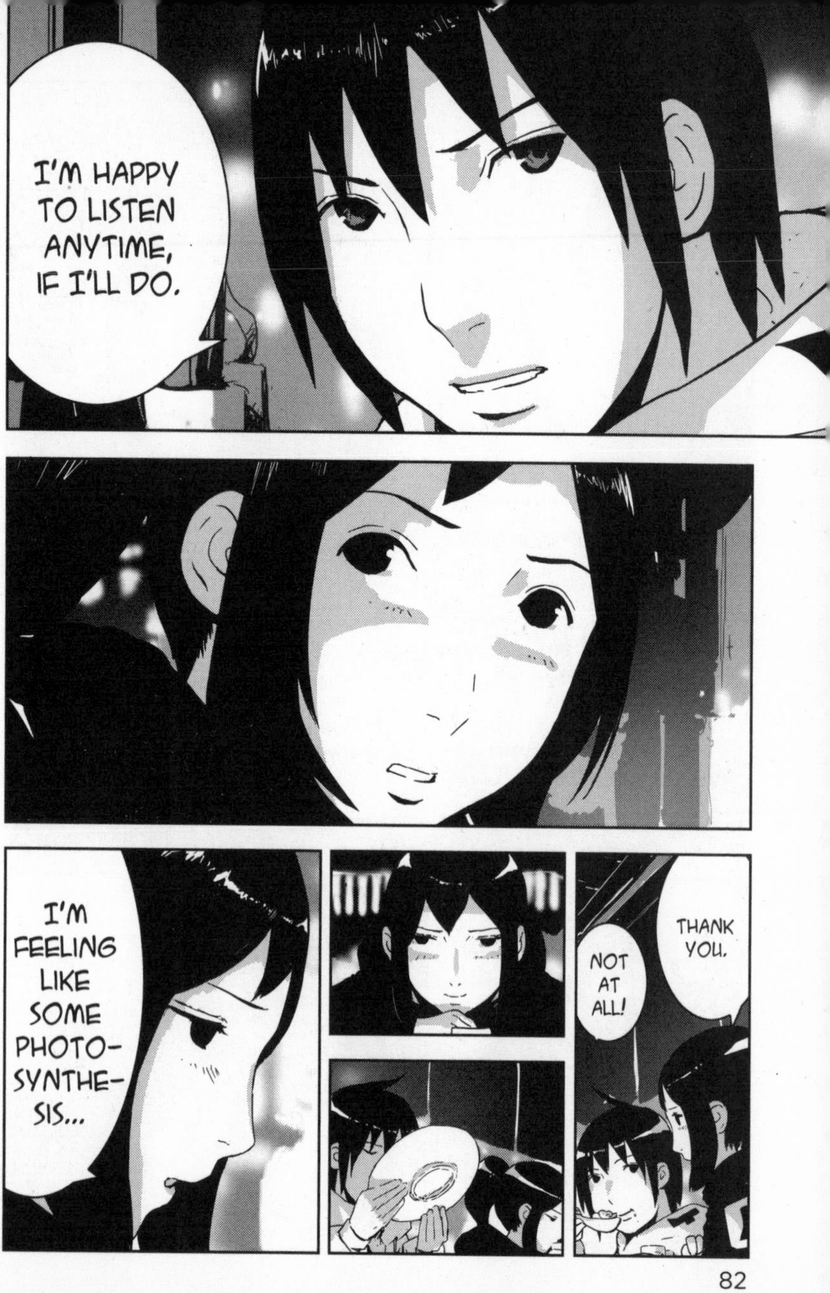
I'M HAPPY TO LISTEN ANYTIME, IF I'LL DO.
THANK YOU.
NOT AT ALL!
I'M FEELING LIKE SOME PHOTO-SYNTHE-SIS...

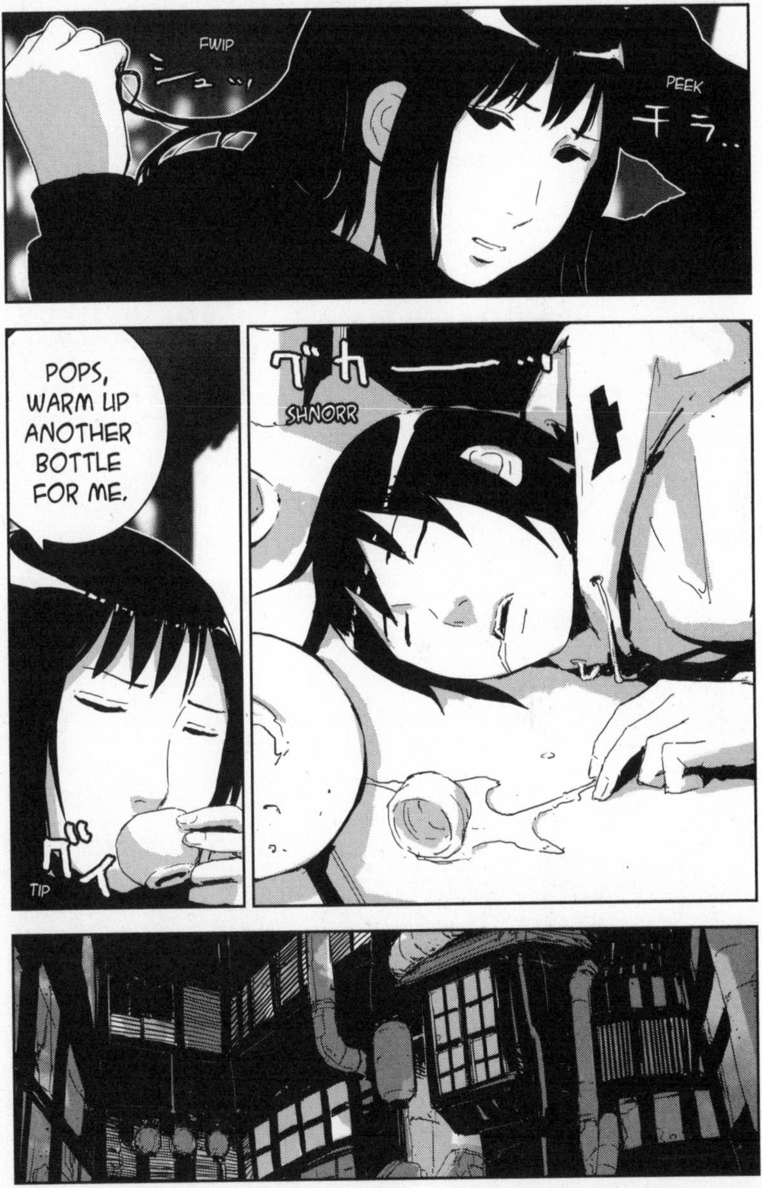
PEEK
チラ…
FWIP
シュ…ッ
ぐか——…
SHNORR
POPS, WARM UP ANOTHER BOTTLE FOR ME.
クイッ
TIP

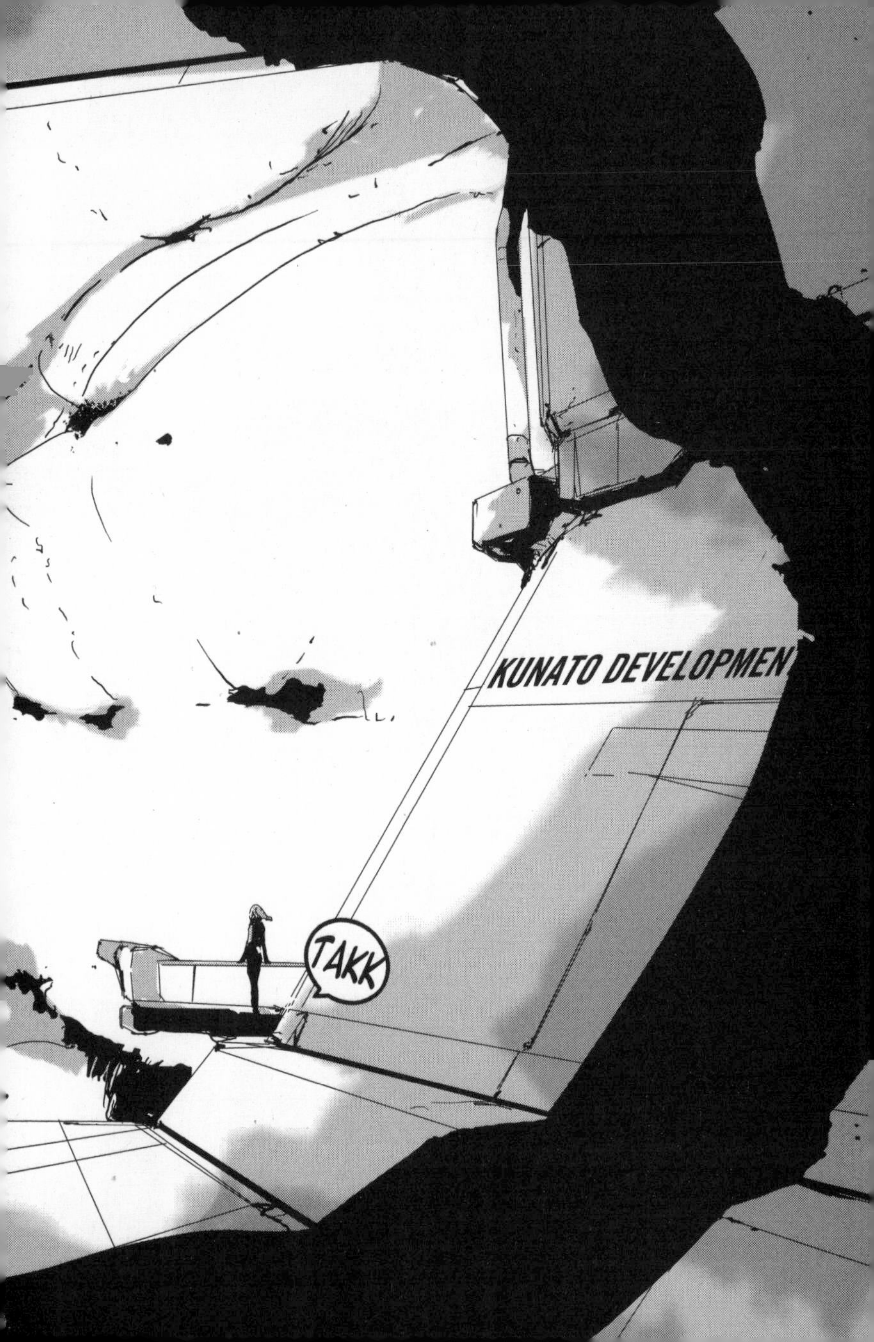
KUNATO DEVELOPMEN
TAKK

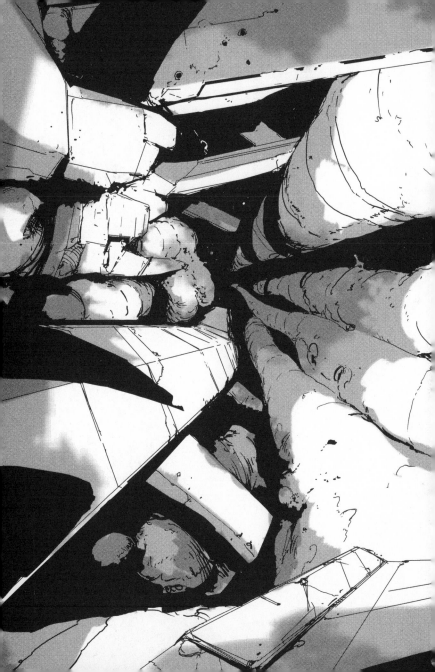

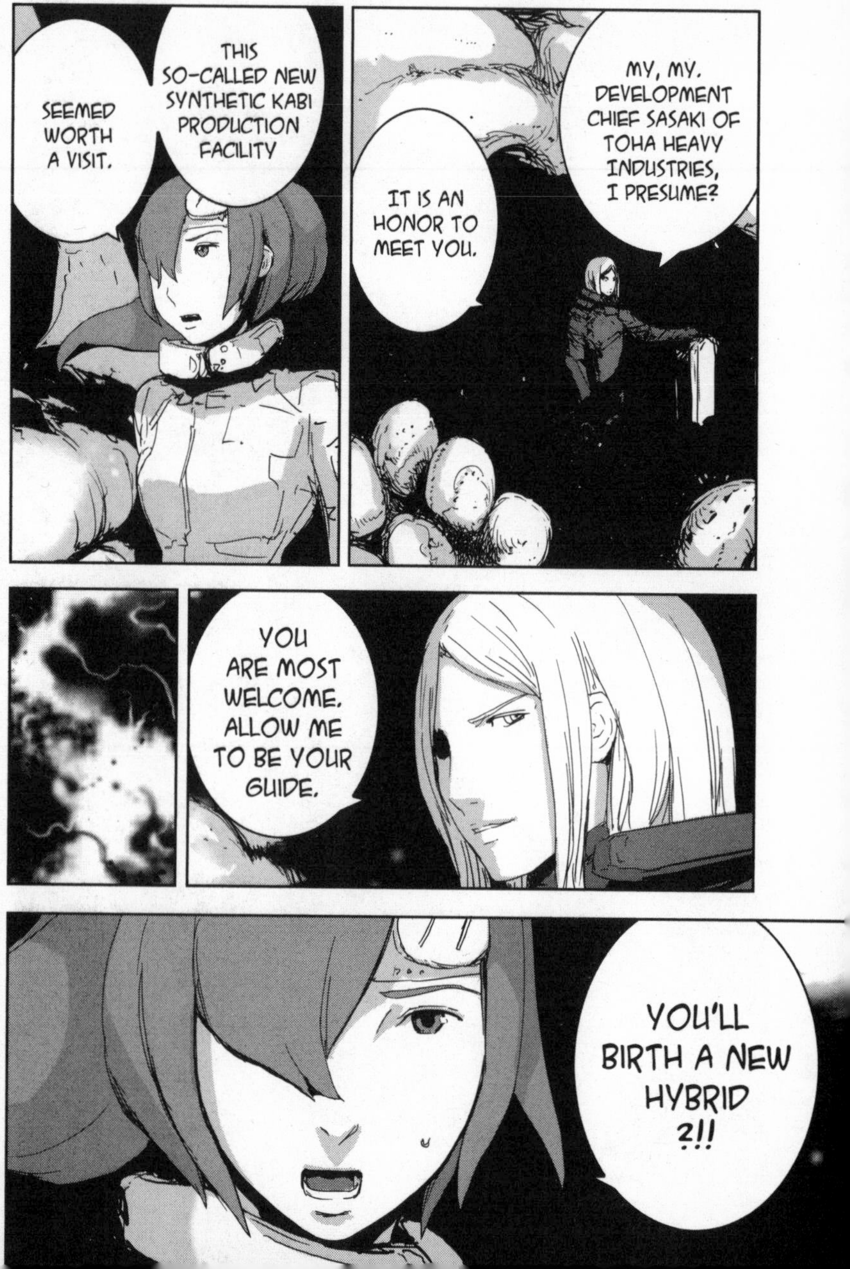

.. ....
MY, MY. DEVELOPMENT CHIEF SASAKI OF TOHA HEAVY INDUSTRIES, I PRESUME?
IT IS AN HONOR TO MEET YOU.
THIS SO-CALLED NEW SYNTHETIC KABI PRODUCTION FACILITY
SEEMED WORTH A VISIT.
YOU ARE MOST WELCOME. ALLOW ME TO BE YOUR GUIDE.
YOU'LL BIRTH A NEW HYBRID ?!!

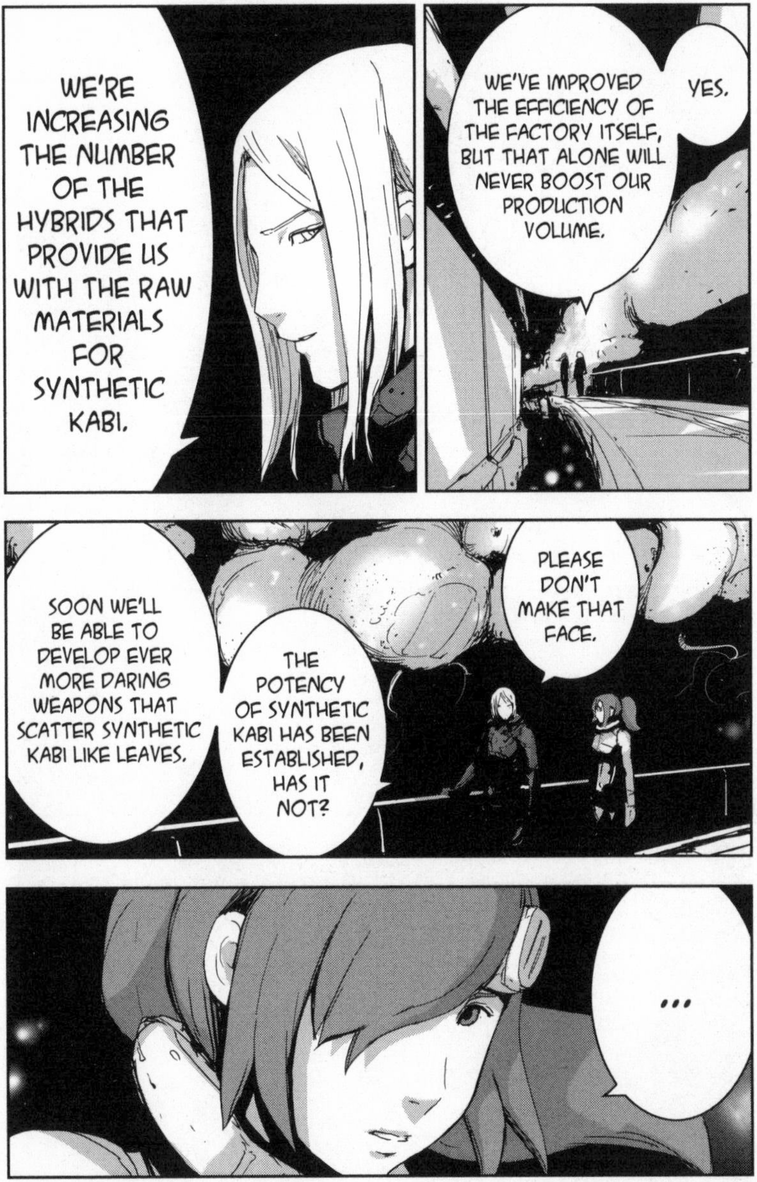
YES.
WE'VE IMPROVED THE EFFICIENCY OF THE FACTORY ITSELF, BUT THAT ALONE WILL NEVER BOOST OUR PRODUCTION VOLUME.
WE'RE INCREASING THE NUMBER OF THE HYBRIDS THAT PROVIDE US WITH THE RAW MATERIALS FOR SYNTHETIC KABI.
PLEASE DON'T MAKE THAT FACE.
THE POTENCY OF SYNTHETIC KABI HAS BEEN ESTABLISHED, HAS IT NOT?
SOON WE'LL BE ABLE TO DEVELOP EVER MORE DARING WEAPONS THAT SCATTER SYNTHETIC KABI LIKE LEAVES.
...

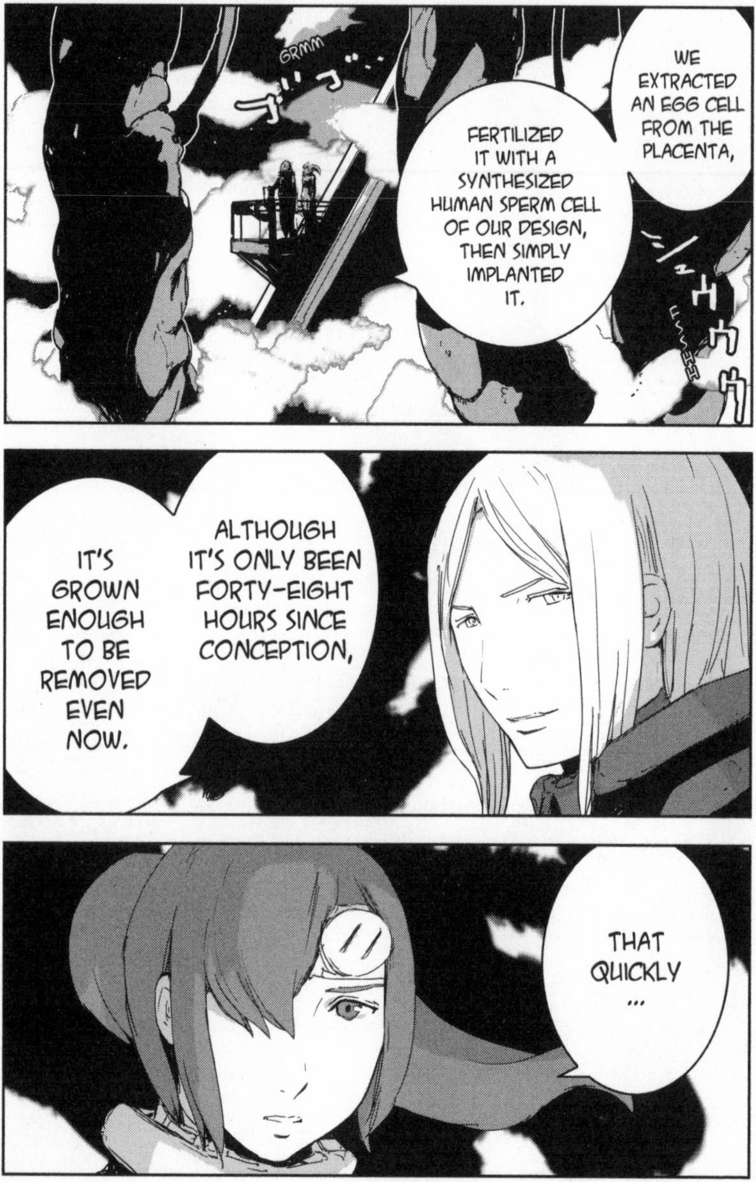
WE EXTRACTED AN EGG CELL FROM THE PLACENTA,
FERTILIZED IT WITH A SYNTHESIZED HUMAN SPERM CELL OF OUR DESIGN, THEN SIMPLY IMPLANTED IT.
GRMM
ALTHOUGH IT'S ONLY BEEN FORTY-EIGHT HOURS SINCE CONCEPTION,
IT'S GROWN ENOUGH TO BE REMOVED EVEN NOW.
THAT QUICKLY ...

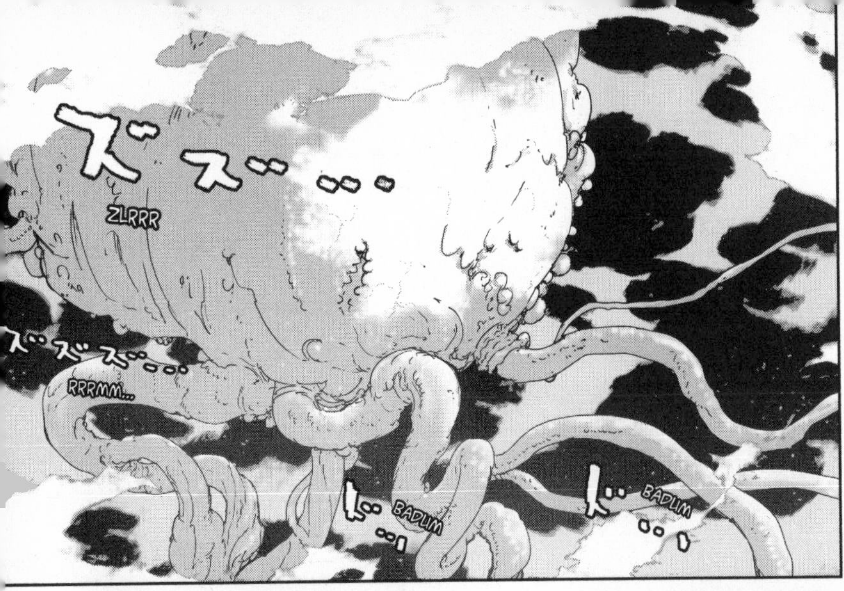
ZLRRR
RRRMM...
BADLIM
BADLIM

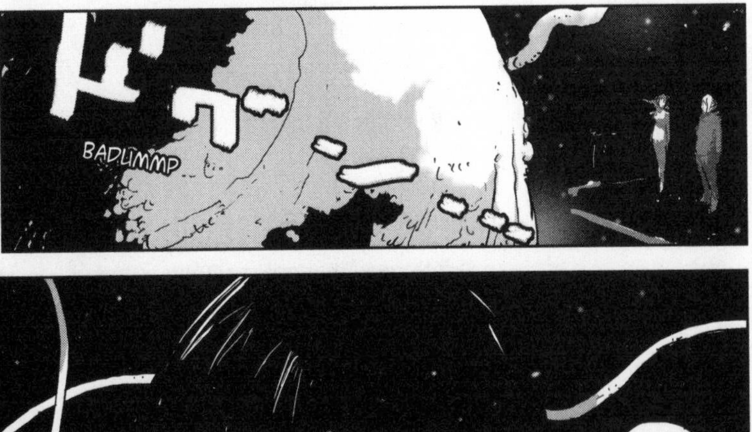
BADLIMMP

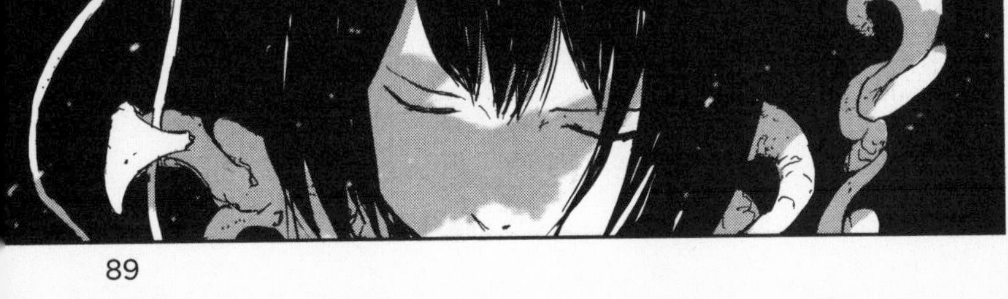

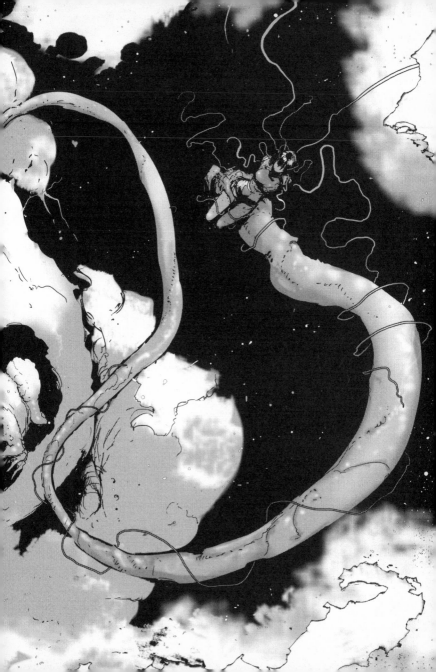

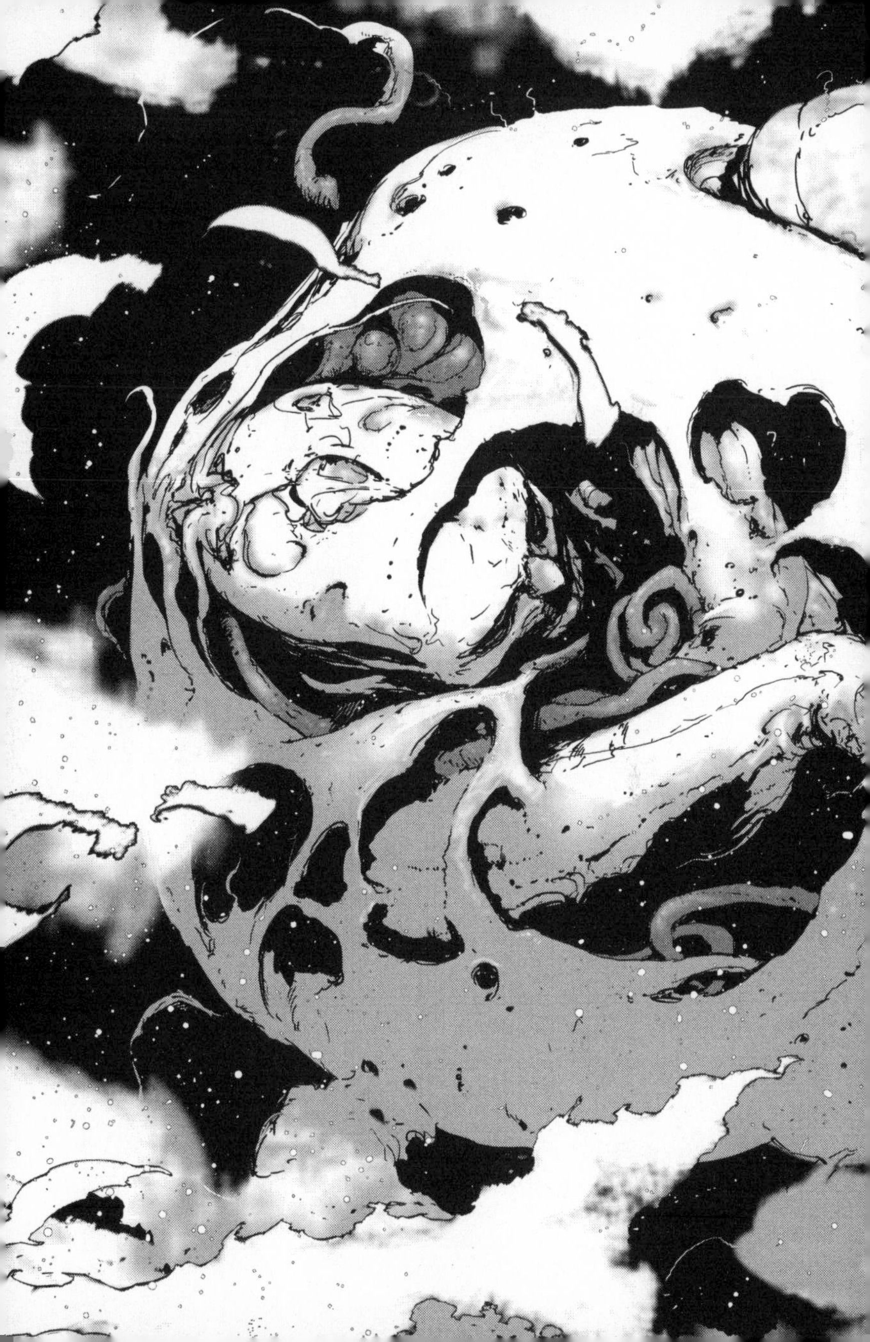

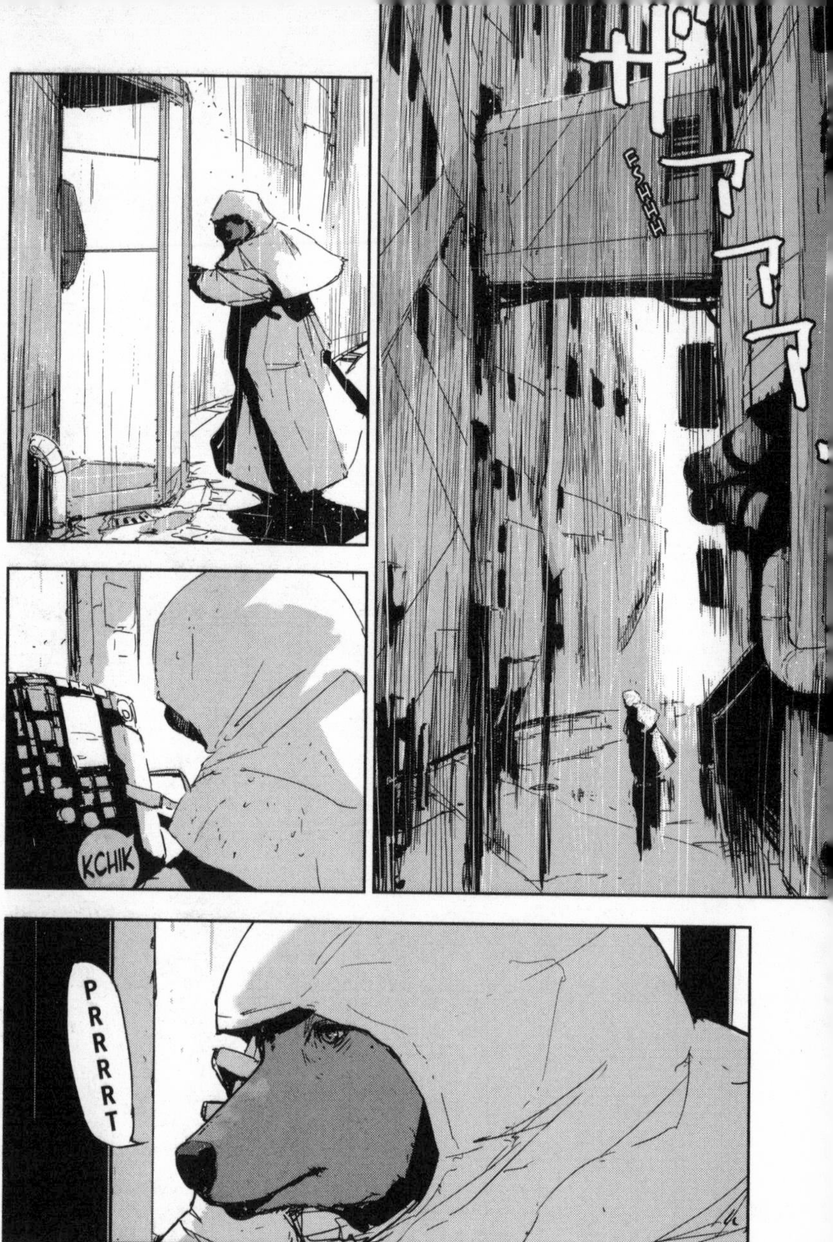
ザァァァ…
WSHHH
KCHIK
PRRRRT

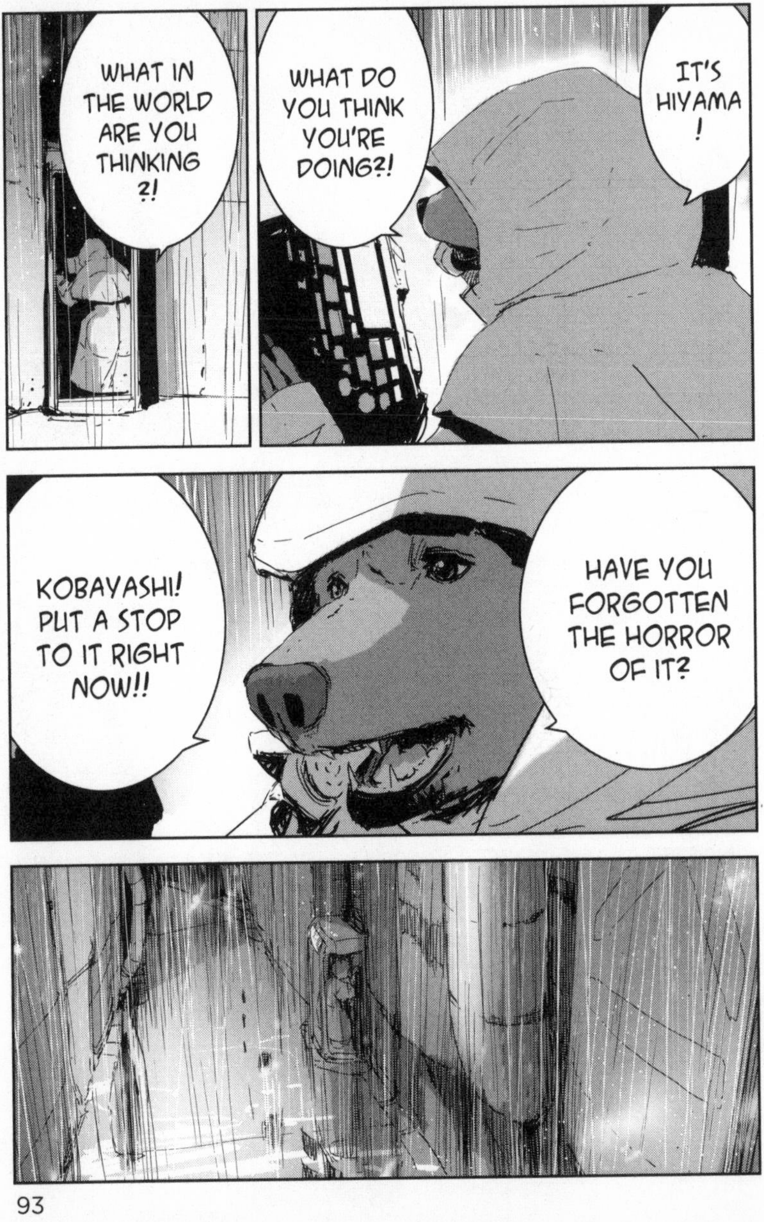
IT'S HIYAMA!
WHAT DO YOU THINK YOU'RE DOING?!
WHAT IN THE WORLD ARE YOU THINKING?!
HAVE YOU FORGOTTEN THE HORROR OF IT?
KOBAYASHI! PUT A STOP TO IT RIGHT NOW!!

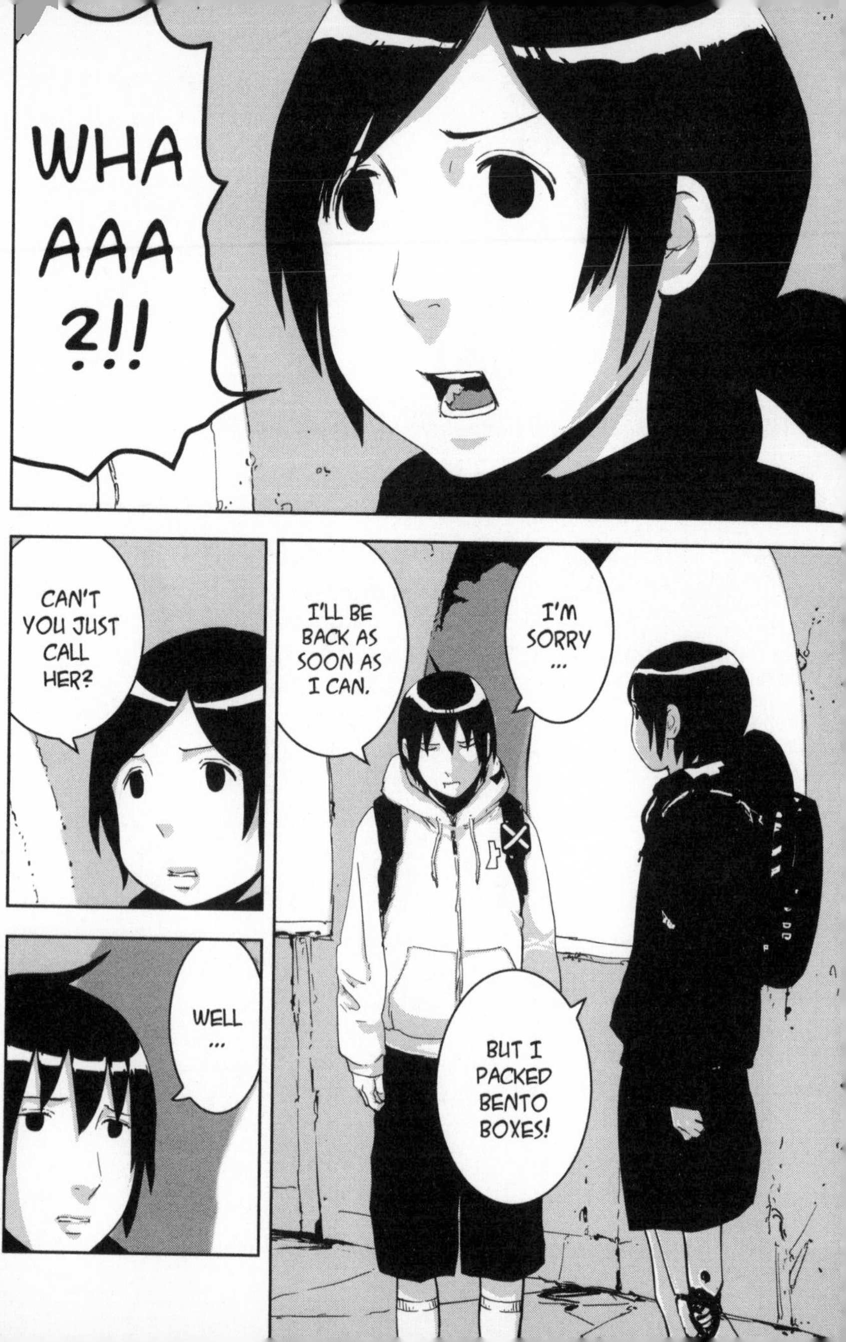
WHA AAA ?!!
I'M SORRY ...
I'LL BE BACK AS SOON AS I CAN.
BUT I PACKED BENTO BOXES!
CAN'T YOU JUST CALL HER?
WELL ...

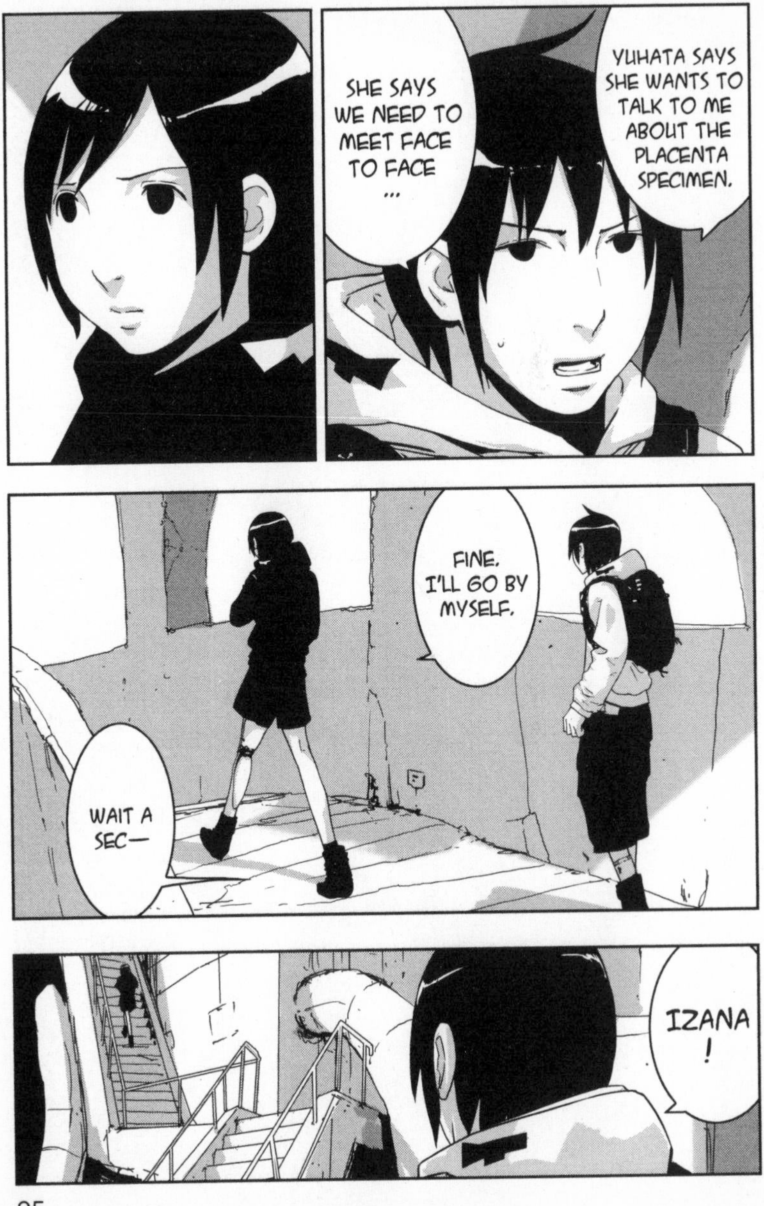
YUHATA SAYS SHE WANTS TO TALK TO ME ABOUT THE PLACENTA SPECIMEN.
SHE SAYS WE NEED TO MEET FACE TO FACE ...
FINE. I'LL GO BY MYSELF.
WAIT A SEC—
IZANA!

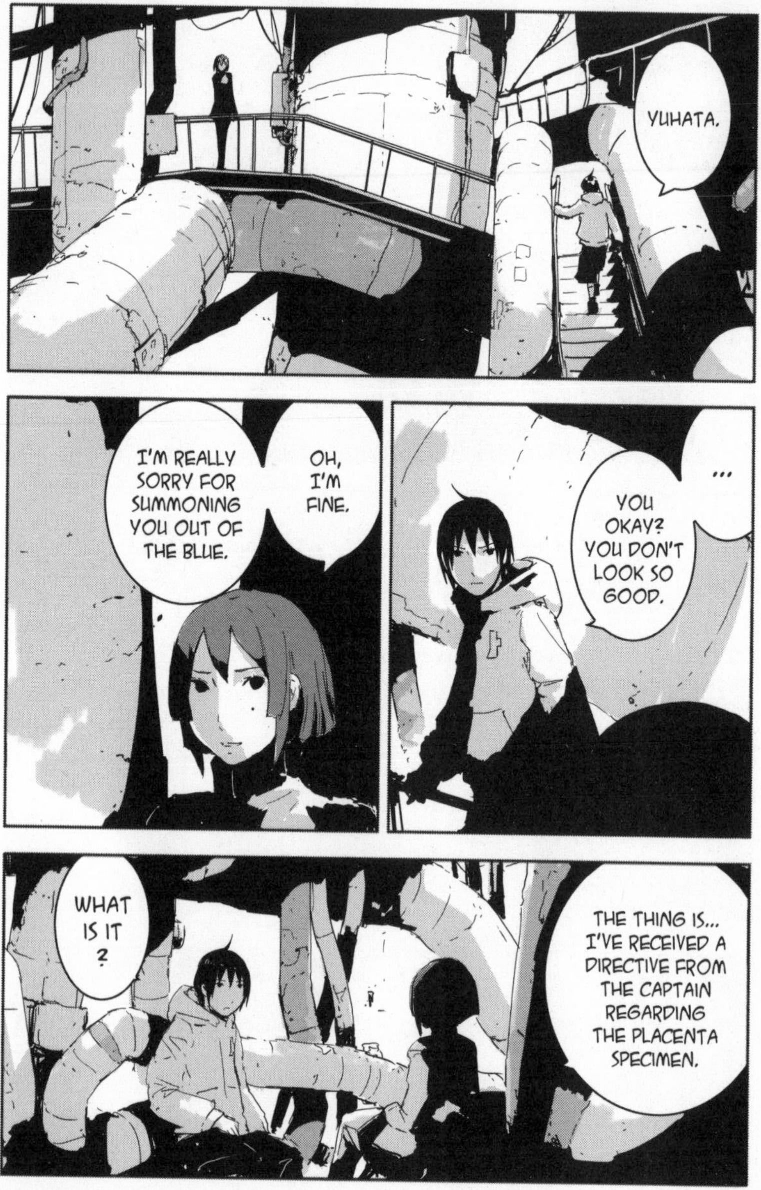
YUHATA.
...
YOU OKAY? YOU DON'T LOOK SO GOOD.
OH, I'M FINE.
I'M REALLY SORRY FOR SUMMONING YOU OUT OF THE BLUE.
WHAT IS IT ?
THE THING IS... I'VE RECEIVED A DIRECTIVE FROM THE CAPTAIN REGARDING THE PLACENTA SPECIMEN.

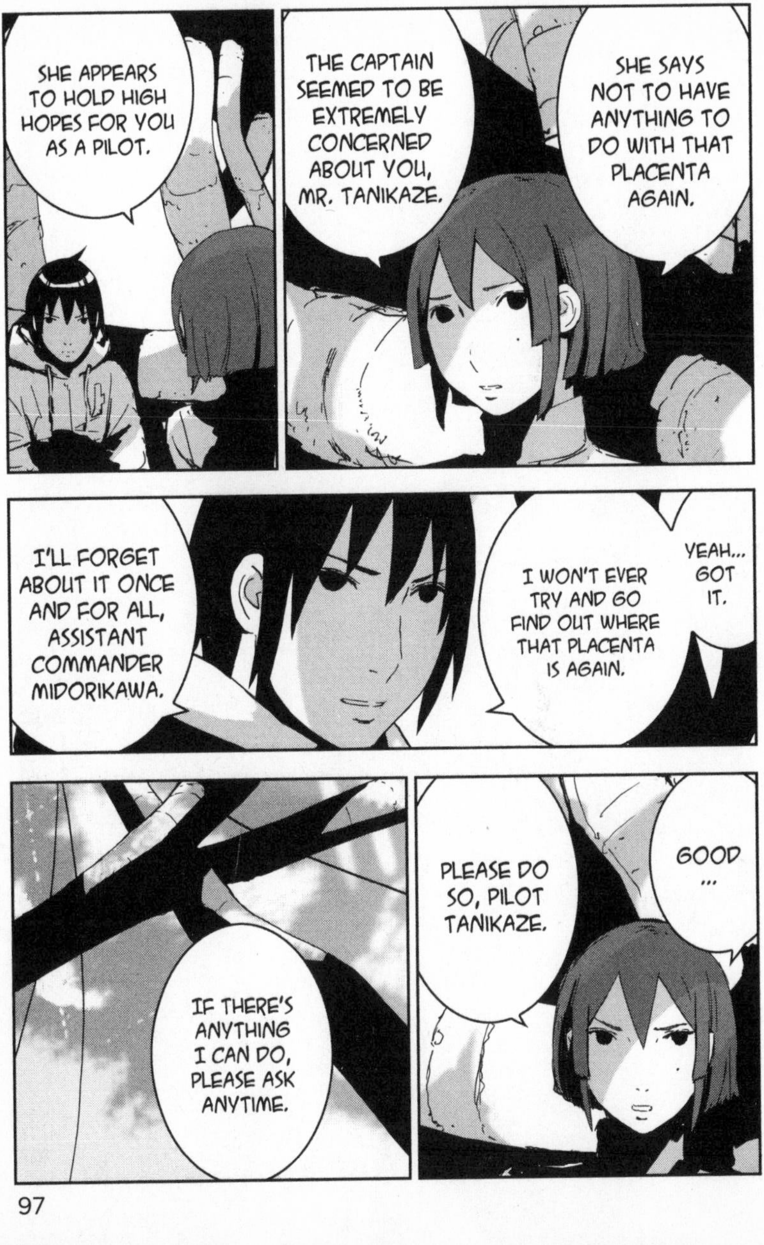
SHE SAYS NOT TO HAVE ANYTHING TO DO WITH THAT PLACENTA AGAIN.
THE CAPTAIN SEEMED TO BE EXTREMELY CONCERNED ABOUT YOU, MR. TANIKAZE.
SHE APPEARS TO HOLD HIGH HOPES FOR YOU AS A PILOT.
YEAH... GOT IT.
I WON'T EVER TRY AND GO FIND OUT WHERE THAT PLACENTA IS AGAIN.
I'LL FORGET ABOUT IT ONCE AND FOR ALL, ASSISTANT COMMANDER MIDORIKAWA.
GOOD ...
PLEASE DO SO, PILOT TANIKAZE.
IF THERE'S ANYTHING I CAN DO, PLEASE ASK ANYTIME.

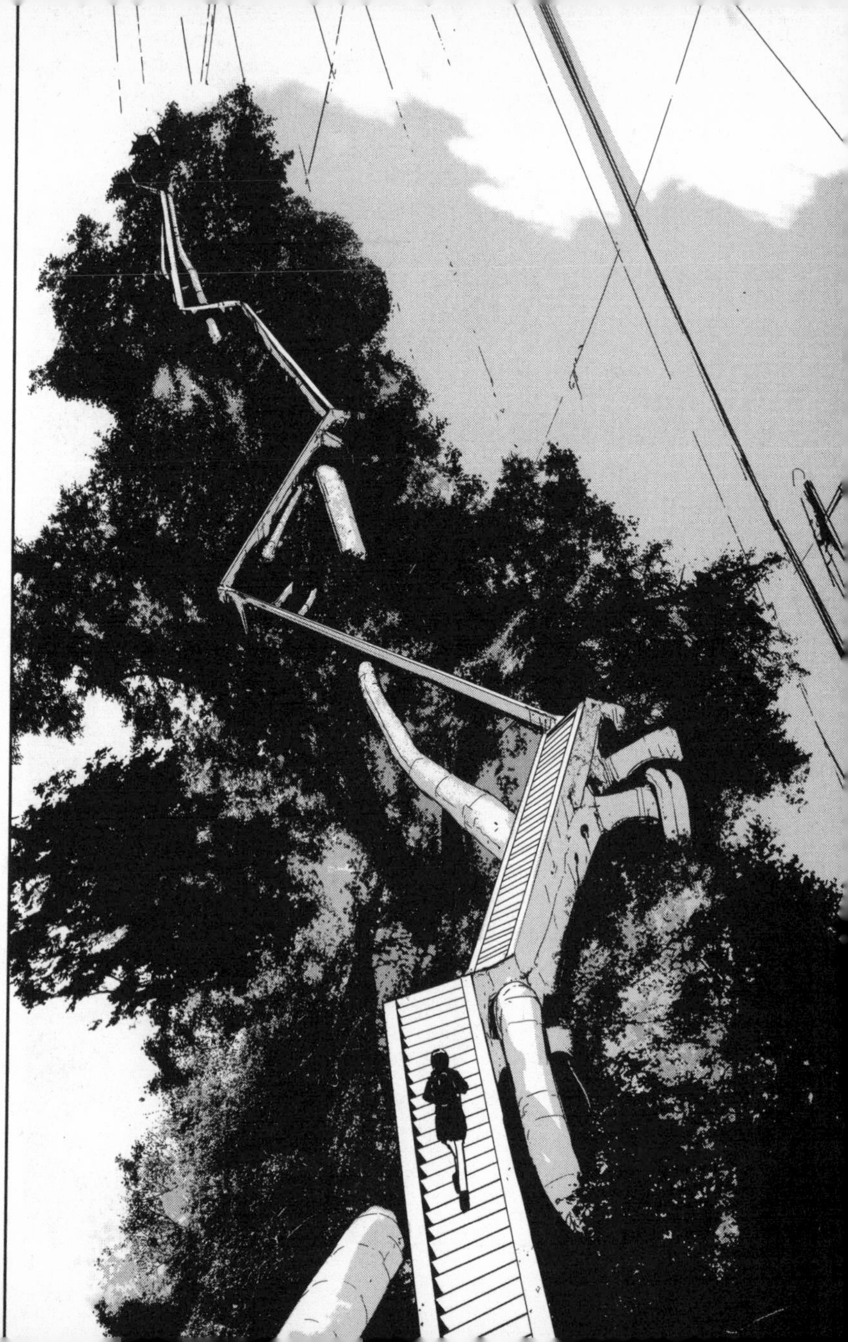

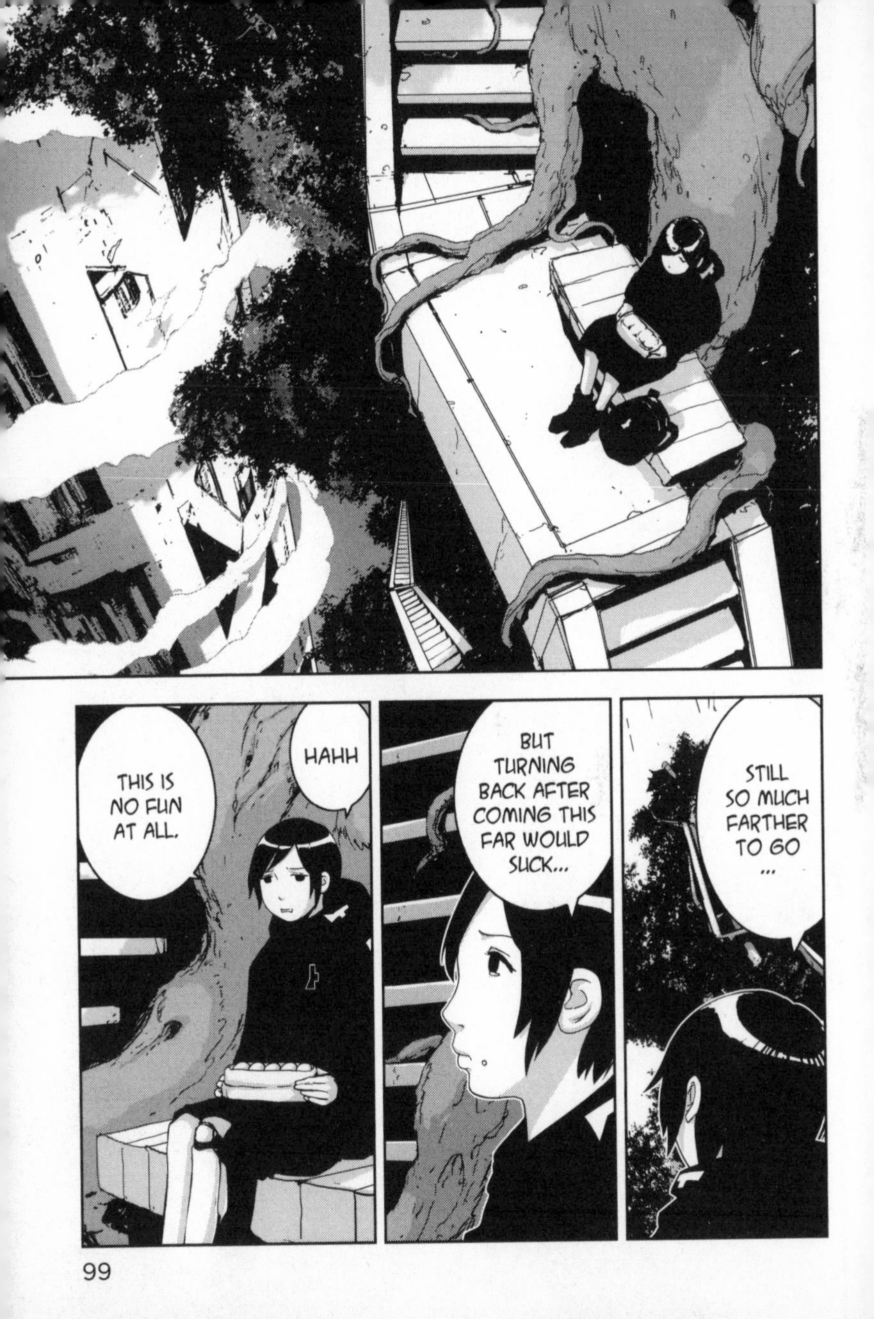
STILL SO MUCH FARTHER TO GO ...
BUT TURNING BACK AFTER COMING THIS FAR WOULD SUCK...
HAHH
THIS IS NO FUN AT ALL.

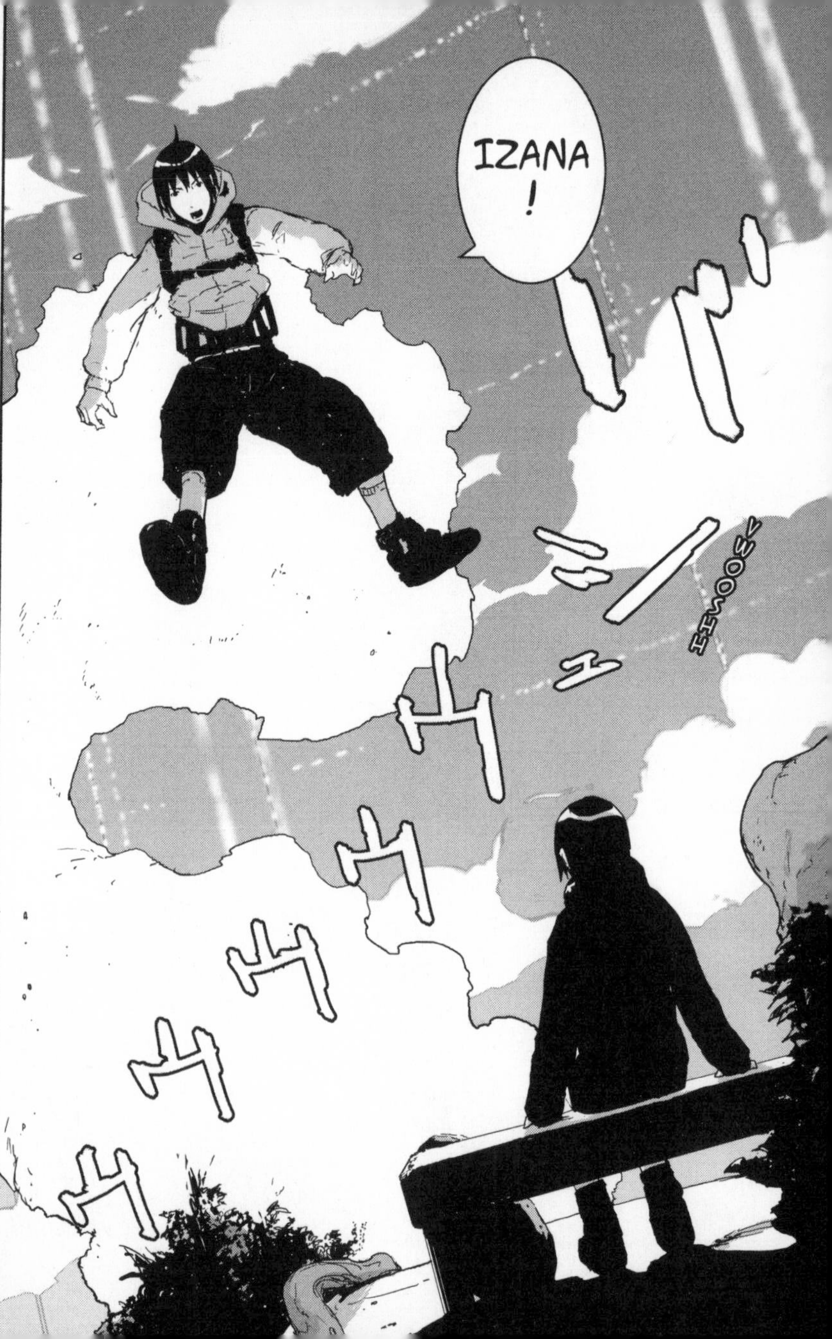
IZANA
!
WOOSH

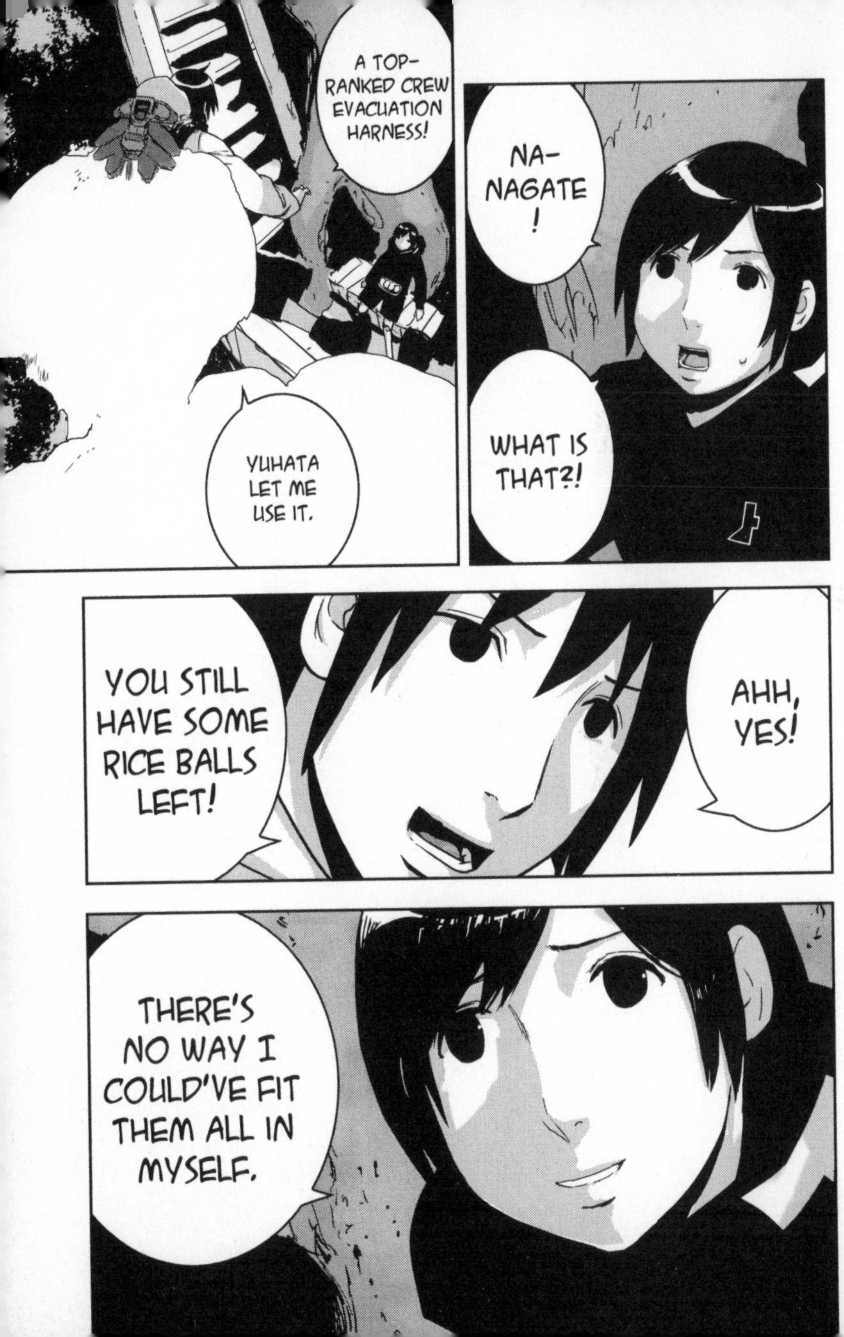
NA-NAGATE!
WHAT IS THAT?!
A TOP-RANKED CREW EVACUATION HARNESS!
YUHATA LET ME USE IT.
AHH, YES!
YOU STILL HAVE SOME RICE BALLS LEFT!
THERE'S NO WAY I COULD'VE FIT THEM ALL IN MYSELF.

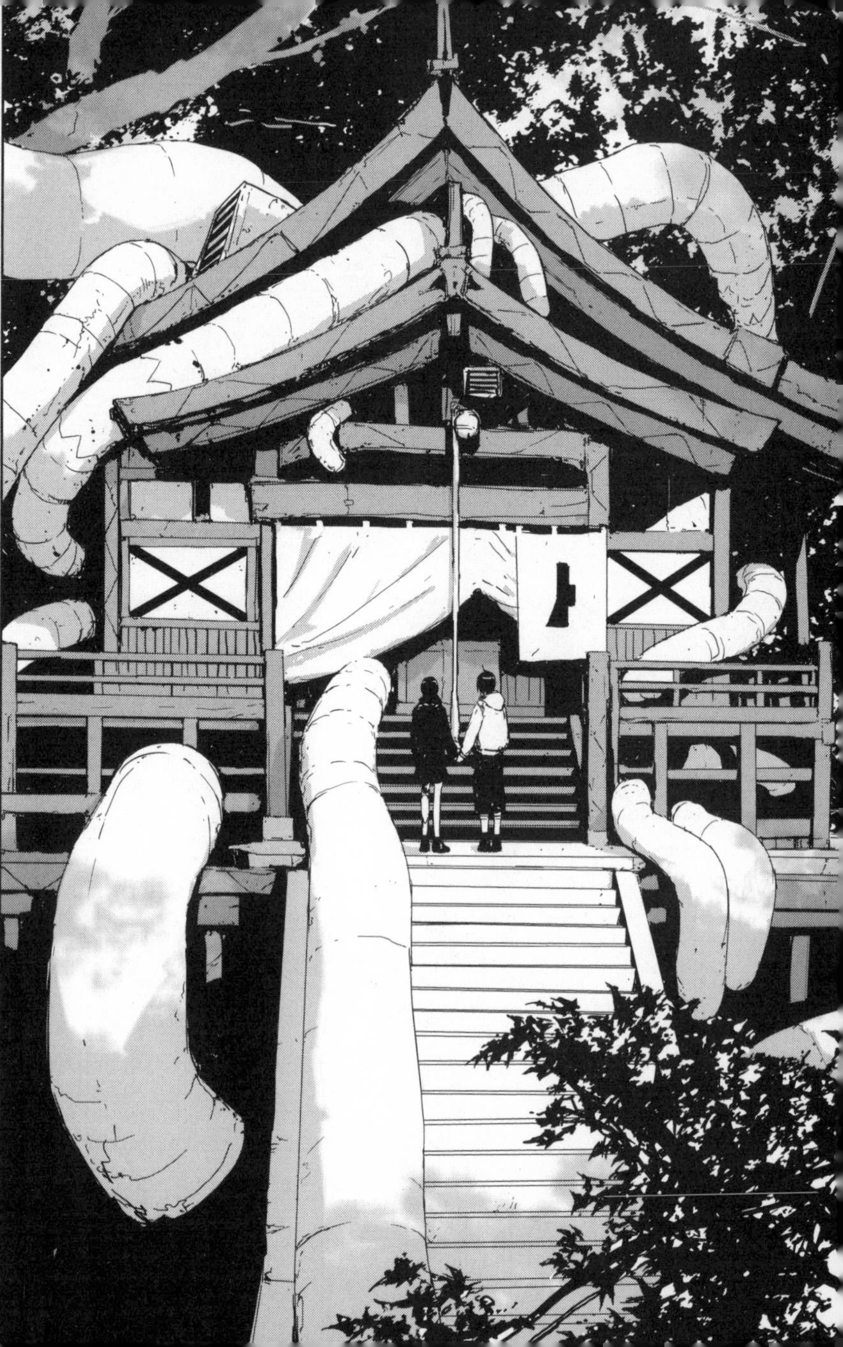

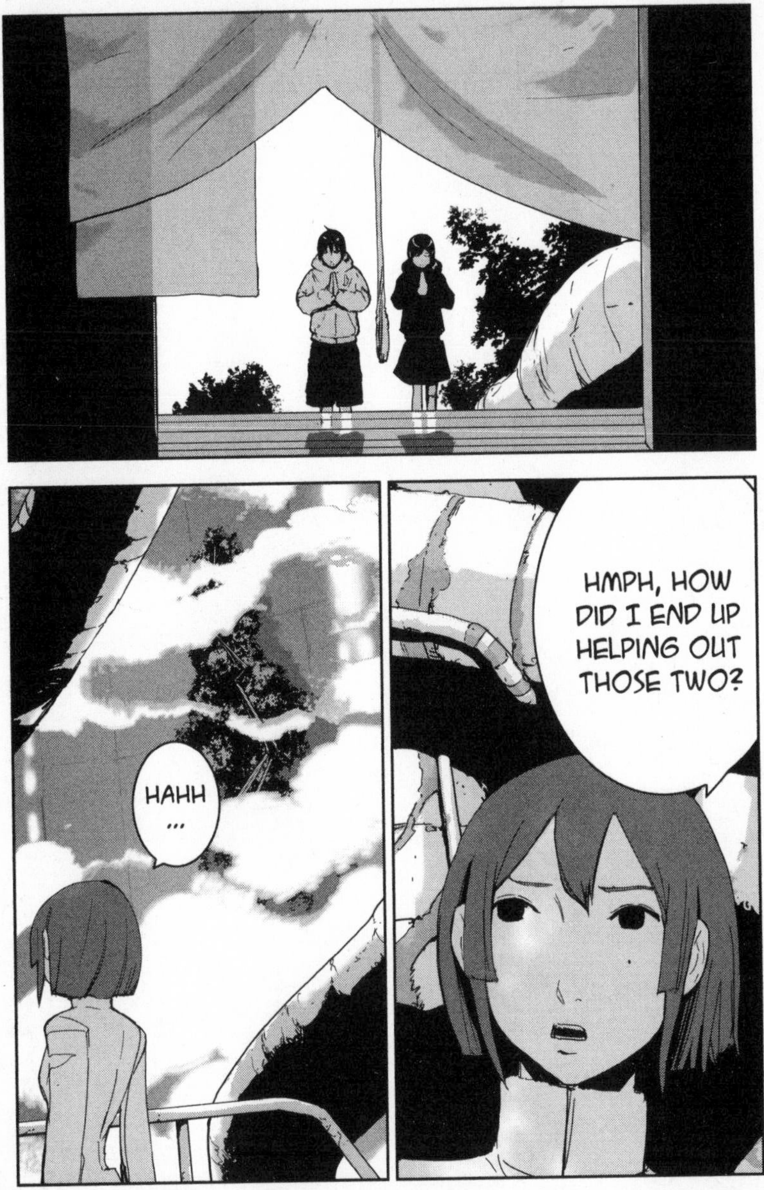
HMPH, HOW DID I END UP HELPING OUT THOSE TWO?
HAHH ...

Chapter 28: END

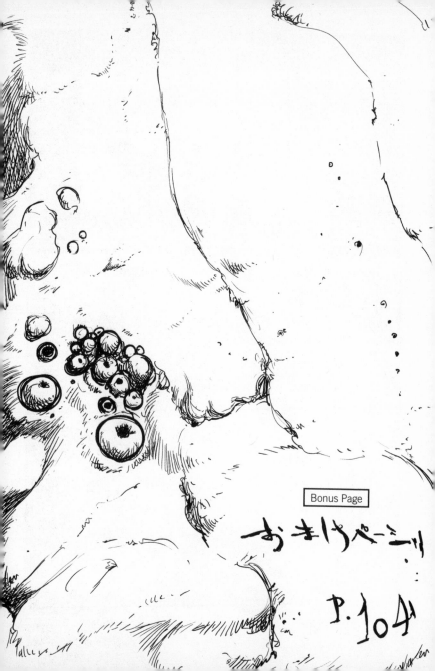

Bonus Page

おまけページ

P.104

Chapter 29: Yure Shinatose's Hesitation

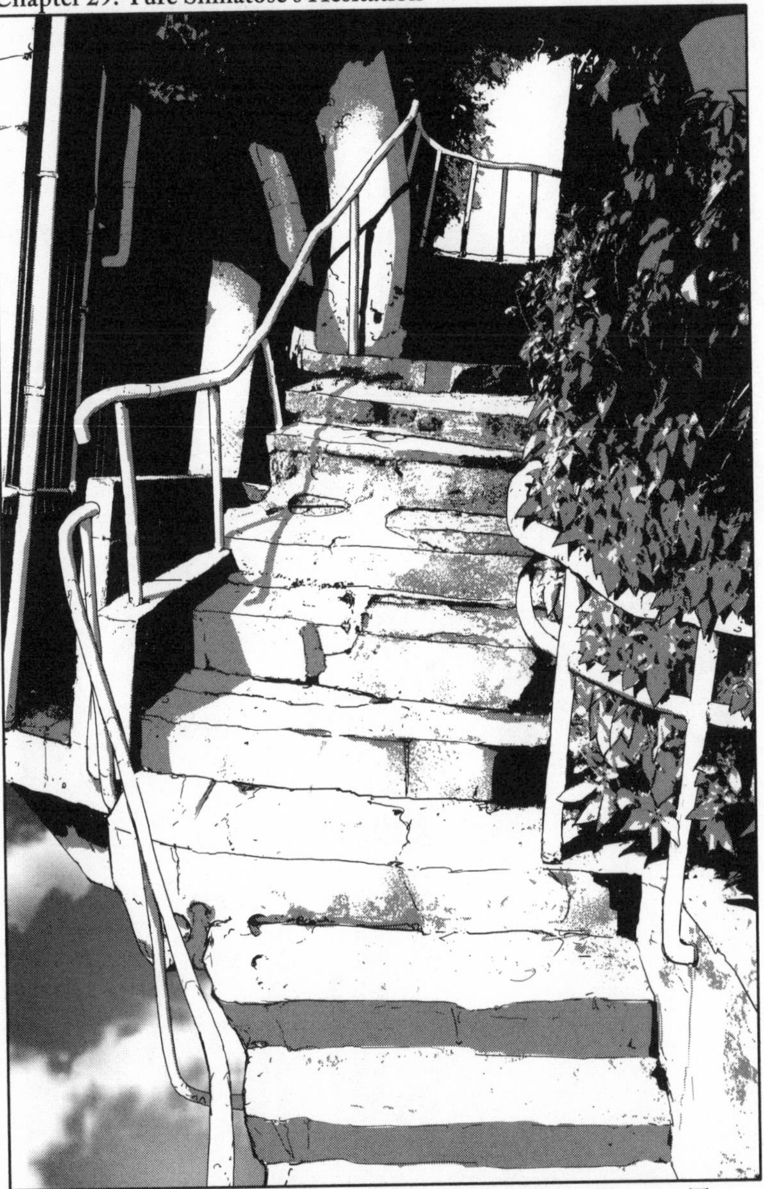

One Hundred Sights of Sidonia Part Twenty-Three:
Pilots' Outer Stairwell

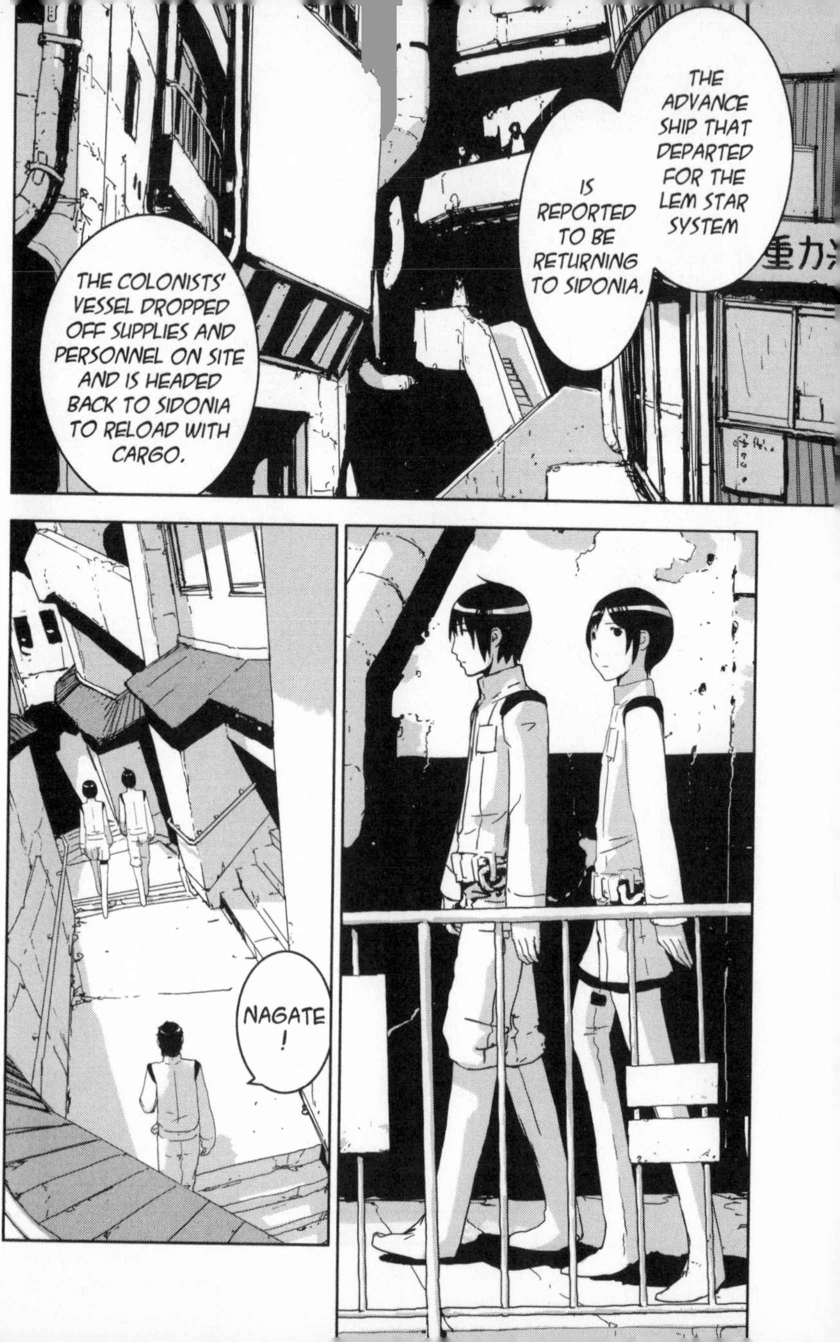
THE ADVANCE SHIP THAT DEPARTED FOR THE LEM STAR SYSTEM
IS REPORTED TO BE RETURNING TO SIDONIA.
THE COLONISTS' VESSEL DROPPED OFF SUPPLIES AND PERSONNEL ON SITE AND IS HEADED BACK TO SIDONIA TO RELOAD WITH CARGO.
NAGATE!

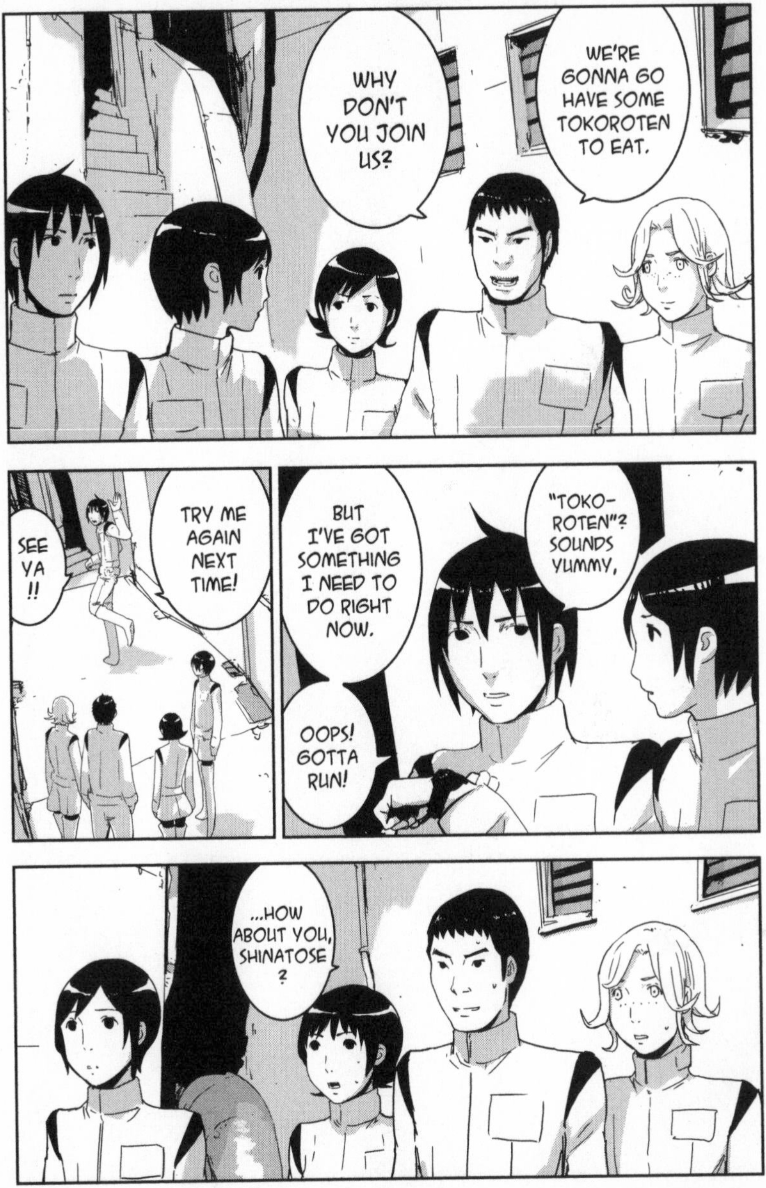
WE'RE GONNA GO HAVE SOME TOKOROTEN TO EAT.
WHY DON'T YOU JOIN US?
"TOKO-ROTEN"? SOUNDS YUMMY,
BUT I'VE GOT SOMETHING I NEED TO DO RIGHT NOW.
OOPS! GOTTA RUN!
TRY ME AGAIN NEXT TIME!
SEE YA!!
...HOW ABOUT YOU, SHINATOSE?

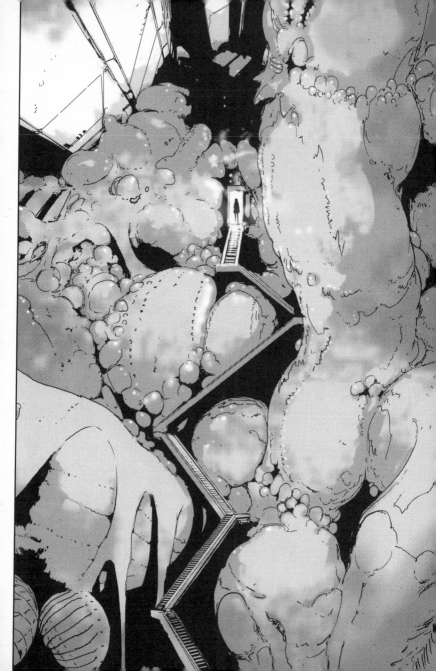

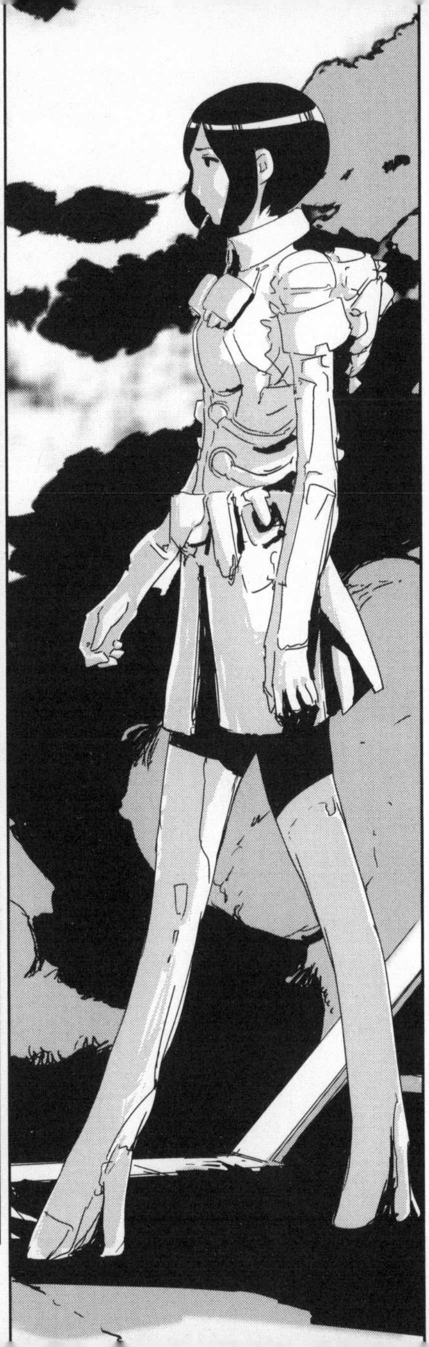

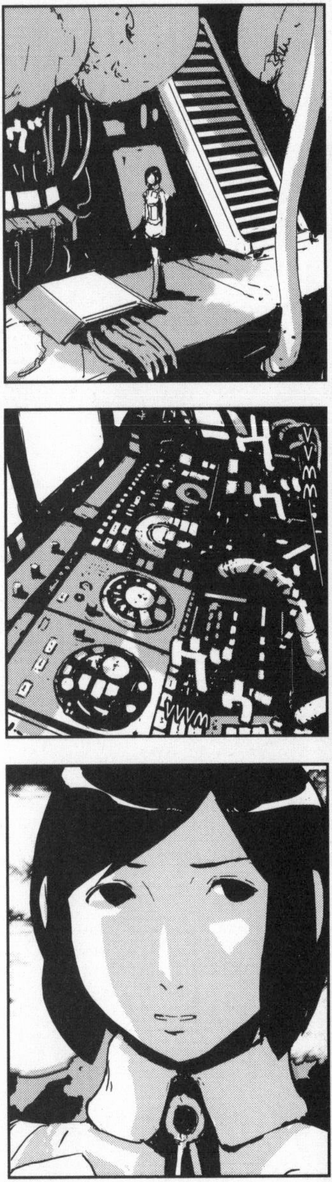

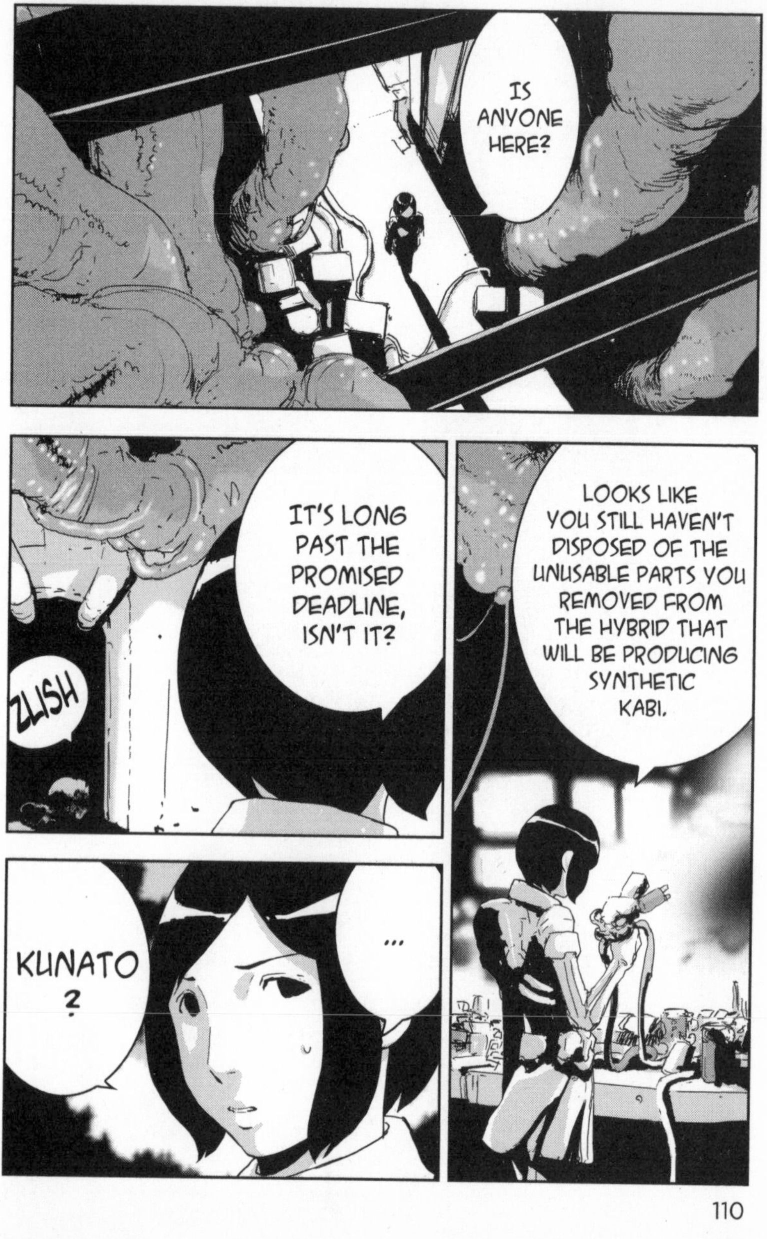
IS ANYONE HERE?
LOOKS LIKE YOU STILL HAVEN'T DISPOSED OF THE UNUSABLE PARTS YOU REMOVED FROM THE HYBRID THAT WILL BE PRODUCING SYNTHETIC KABI.
IT'S LONG PAST THE PROMISED DEADLINE, ISN'T IT?
ZLISH
...
KUNATO?

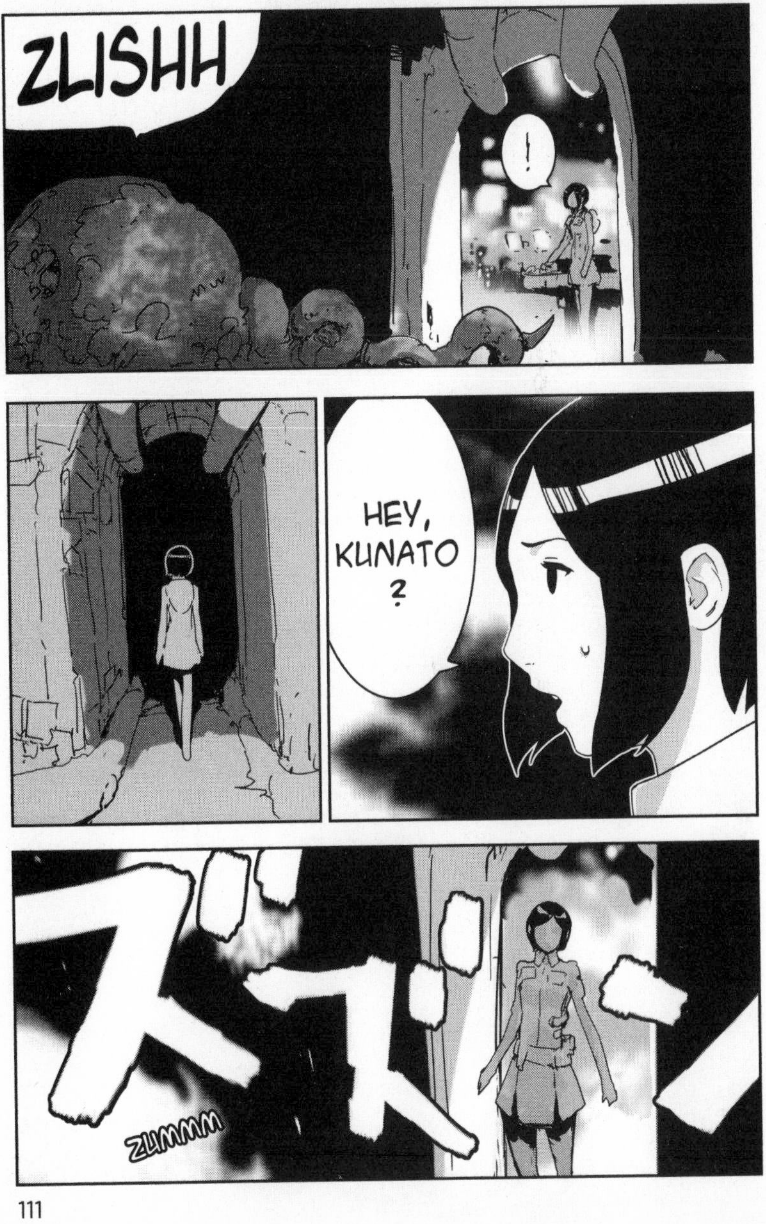

ZLISHH
!
HEY, KUNATO?
ZUMMM

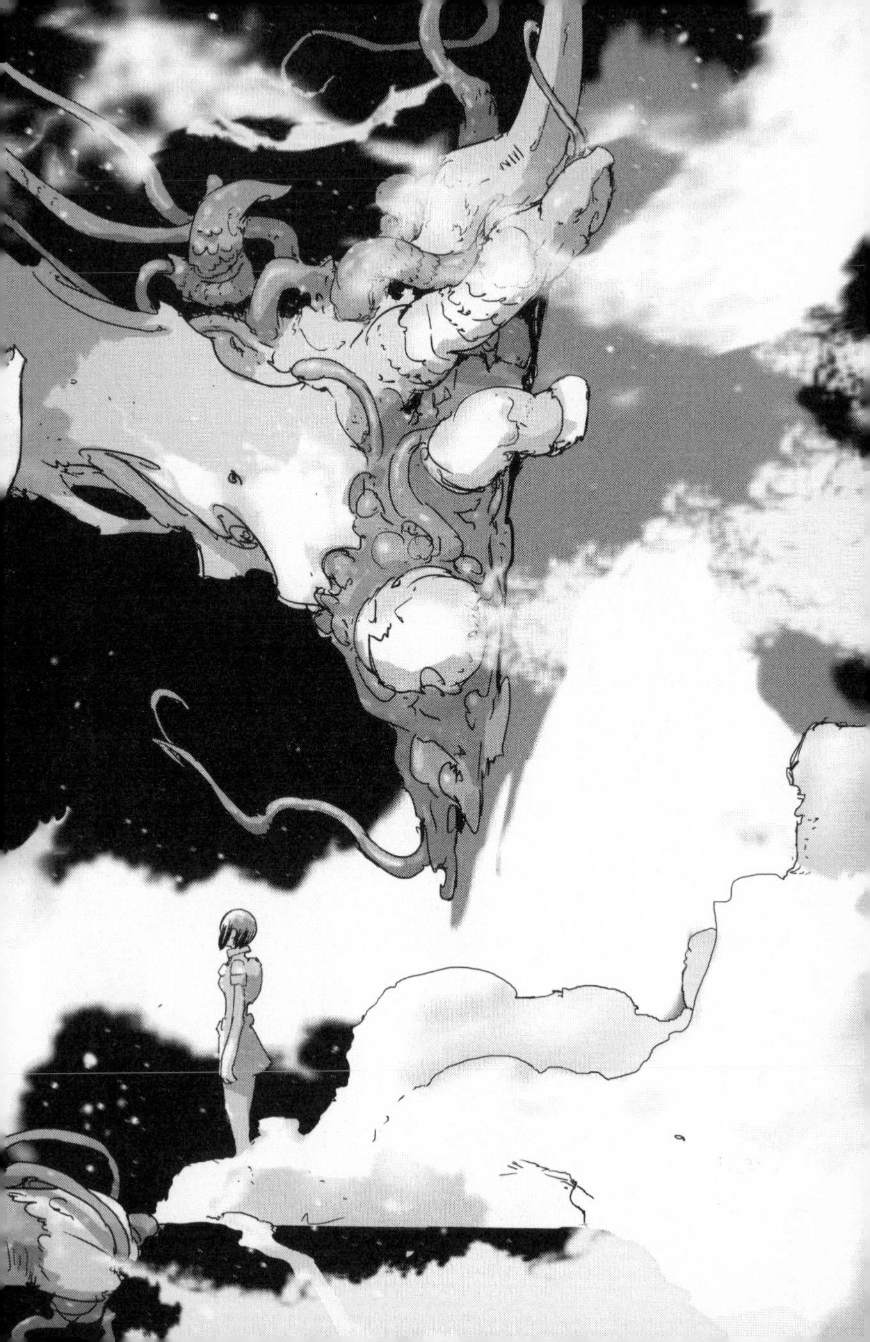

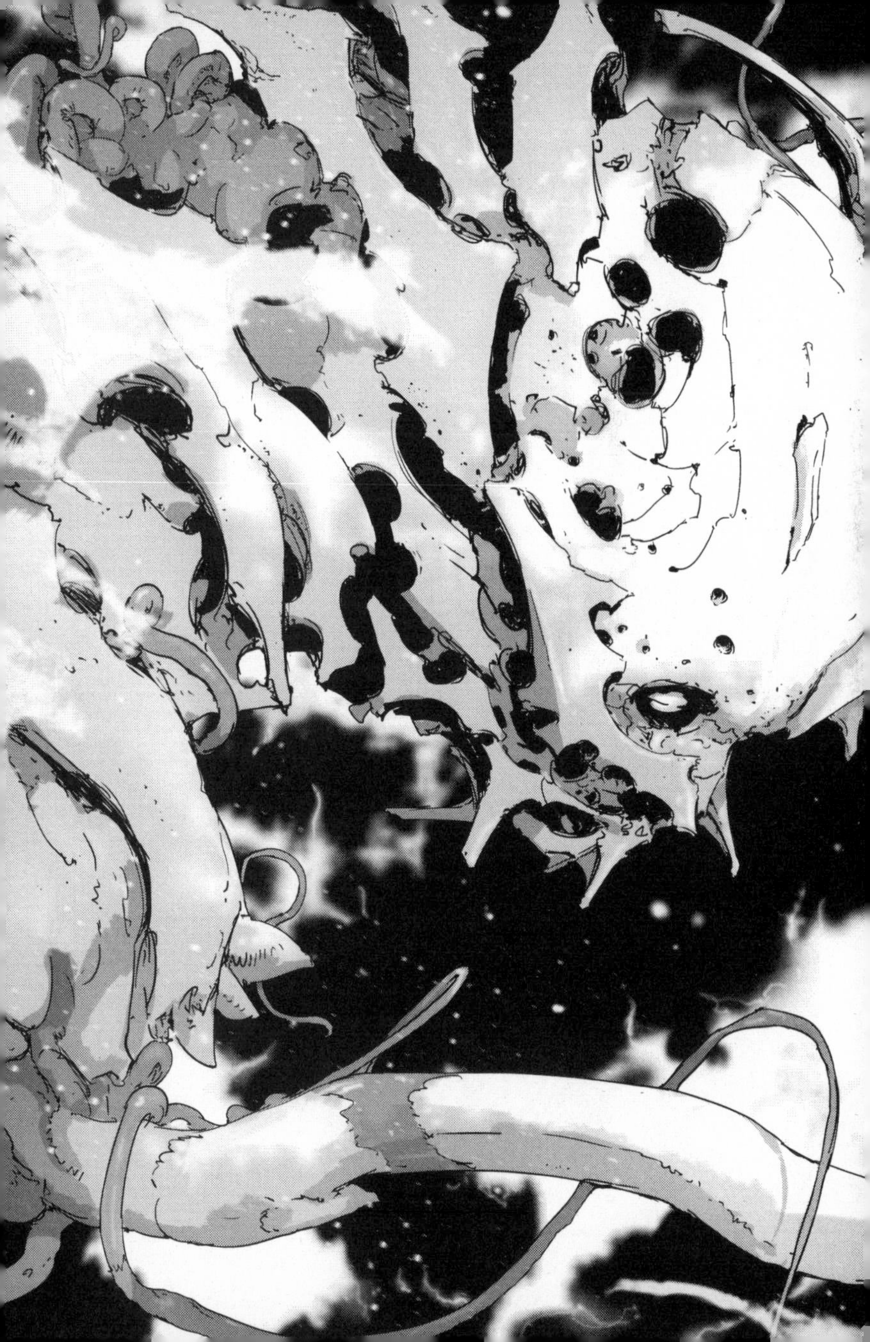

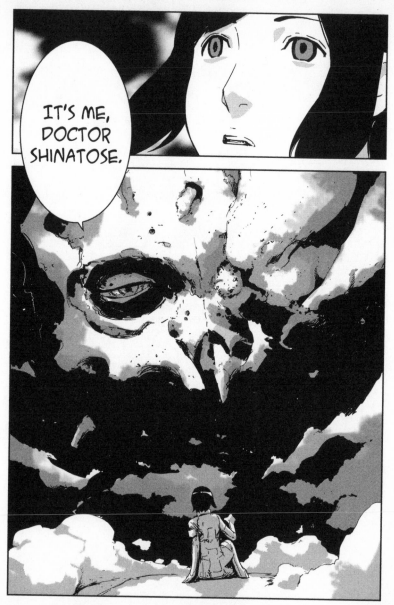

IT'S ME, DOCTOR SHINATOSE.

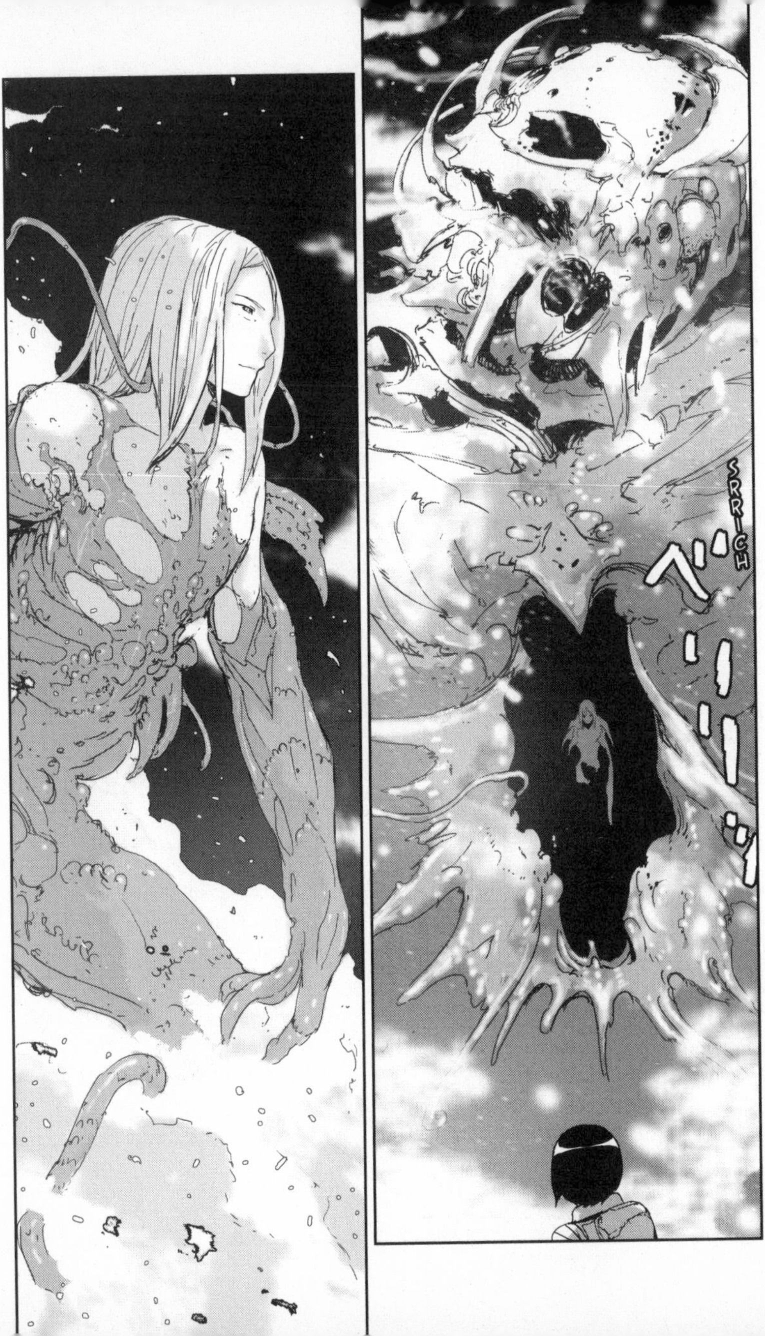
SRRICH

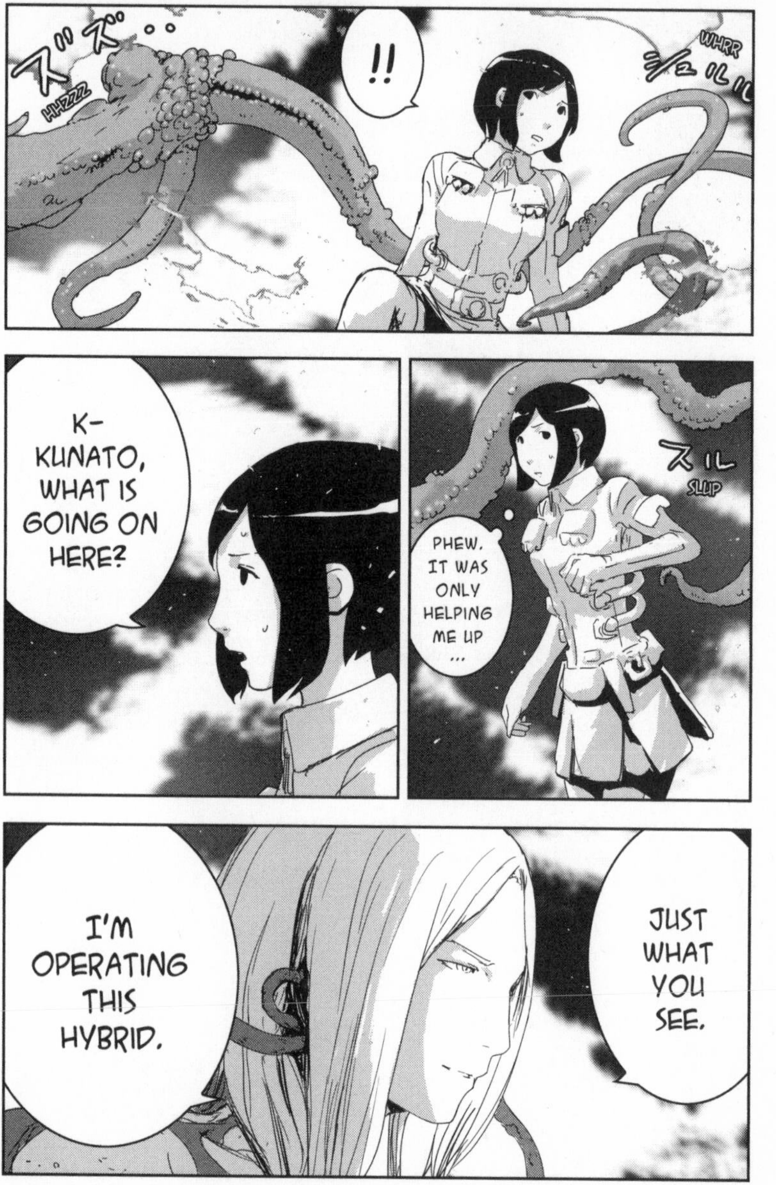
シュルル
WHRR
!!
ズズ..
HHZZZ
スル
SLIP
PHEW. IT WAS ONLY HELPING ME UP ...
K-KUNATO, WHAT IS GOING ON HERE?
JUST WHAT YOU SEE.
I'M OPERATING THIS HYBRID.

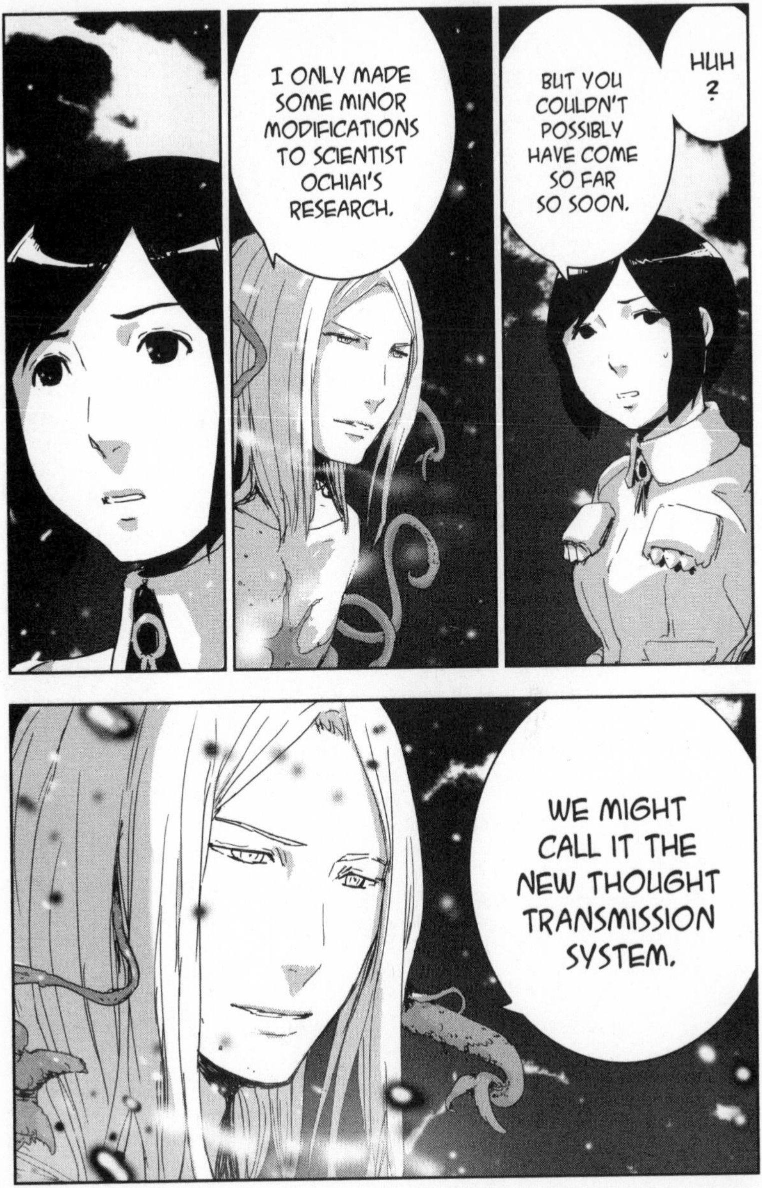
HUH?
BUT YOU COULDN'T POSSIBLY HAVE COME SO FAR SO SOON.
I ONLY MADE SOME MINOR MODIFICATIONS TO SCIENTIST OCHIAI'S RESEARCH.
WE MIGHT CALL IT THE NEW THOUGHT TRANSMISSION SYSTEM.

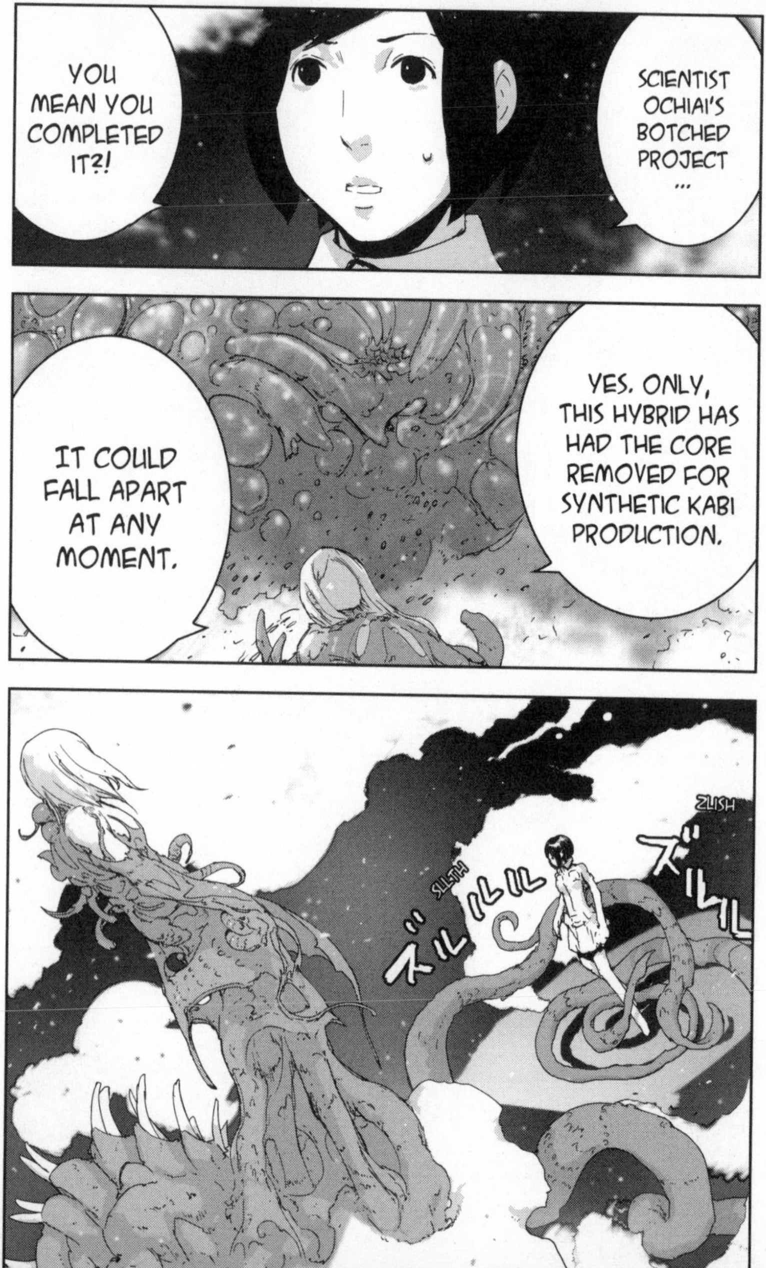
SCIENTIST OCHIAI'S BOTCHED PROJECT ...
YOU MEAN YOU COMPLETED IT?!
YES. ONLY, THIS HYBRID HAS HAD THE CORE REMOVED FOR SYNTHETIC KABI PRODUCTION.
IT COULD FALL APART AT ANY MOMENT.
ZLISH
SLLTH

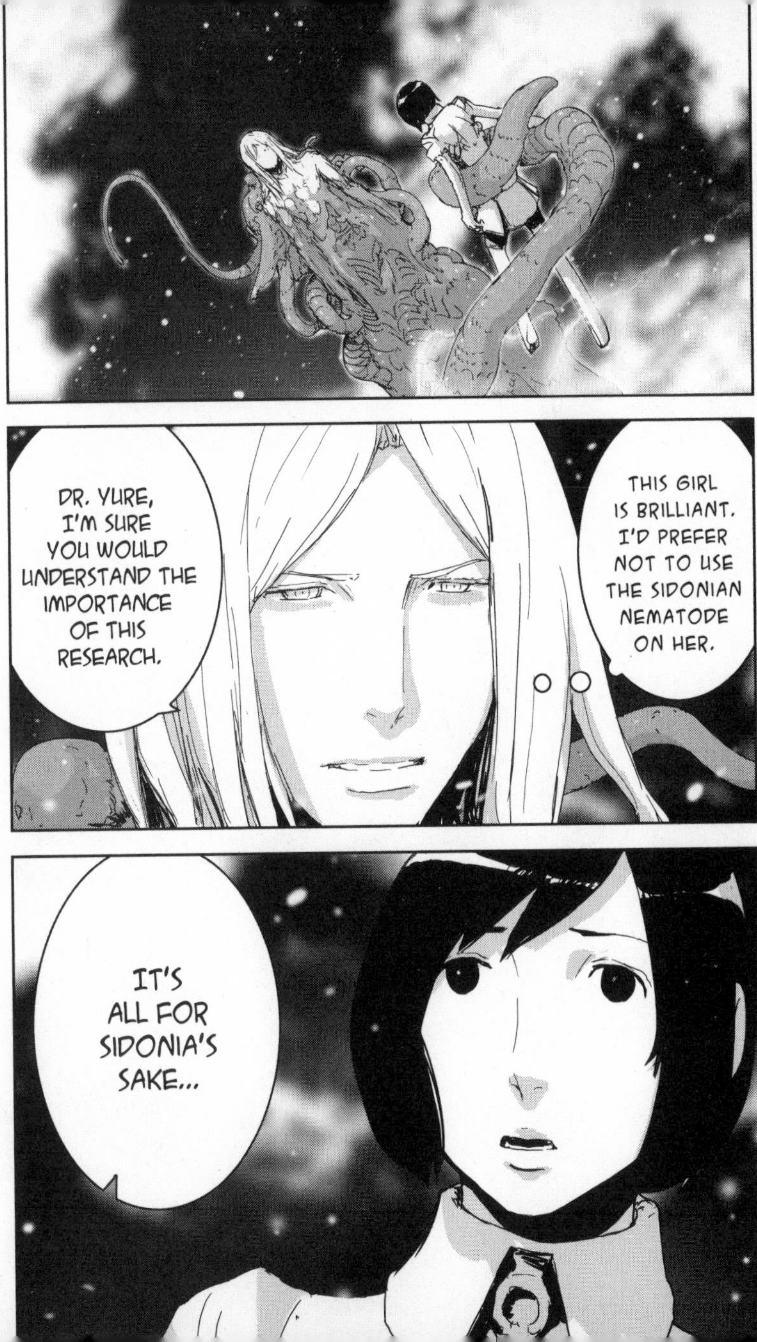
THIS GIRL IS BRILLIANT. I'D PREFER NOT TO USE THE SIDONIAN NEMATODE ON HER.
DR. YURE, I'M SURE YOU WOULD UNDERSTAND THE IMPORTANCE OF THIS RESEARCH.
IT'S ALL FOR SIDONIA'S SAKE...

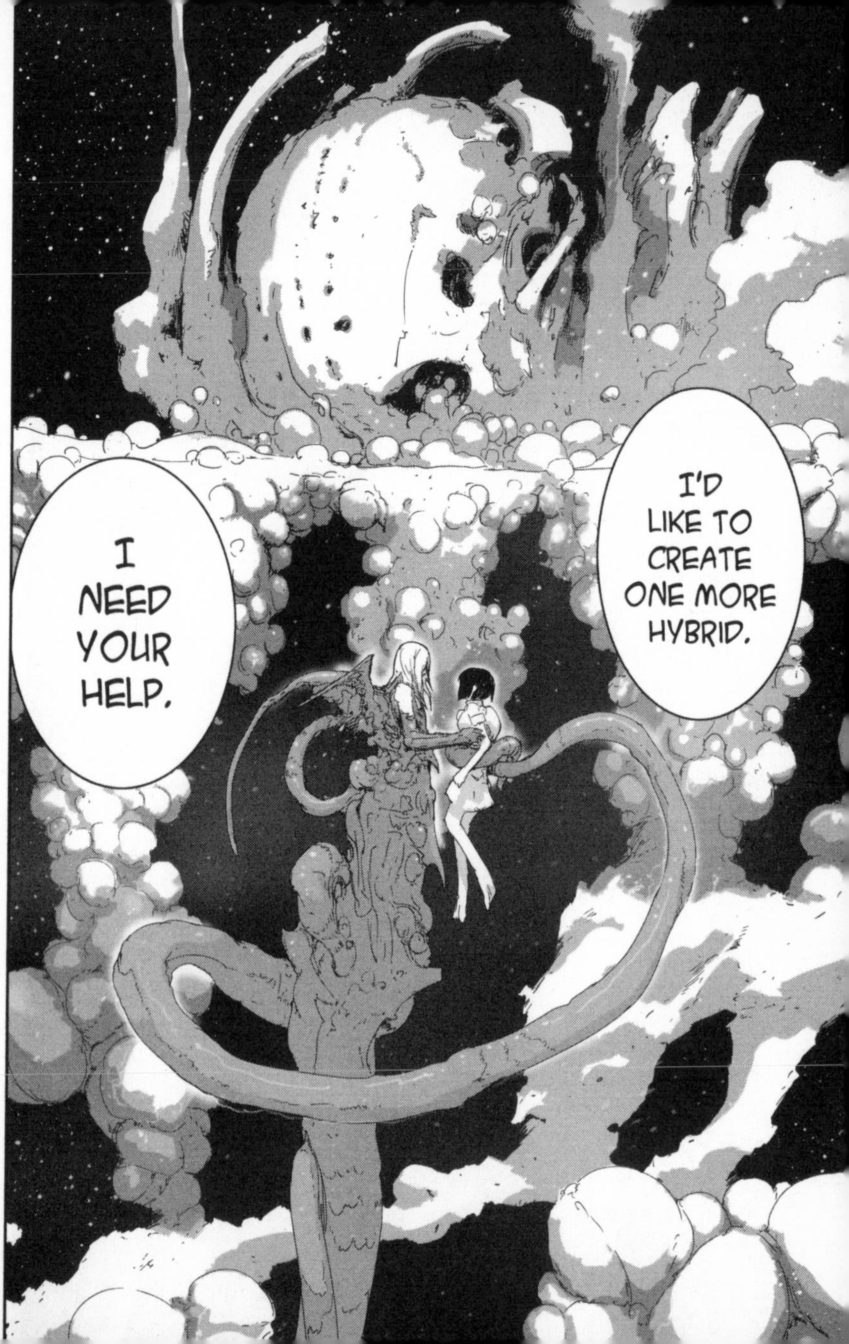
I'D LIKE TO CREATE ONE MORE HYBRID.
I NEED YOUR HELP.

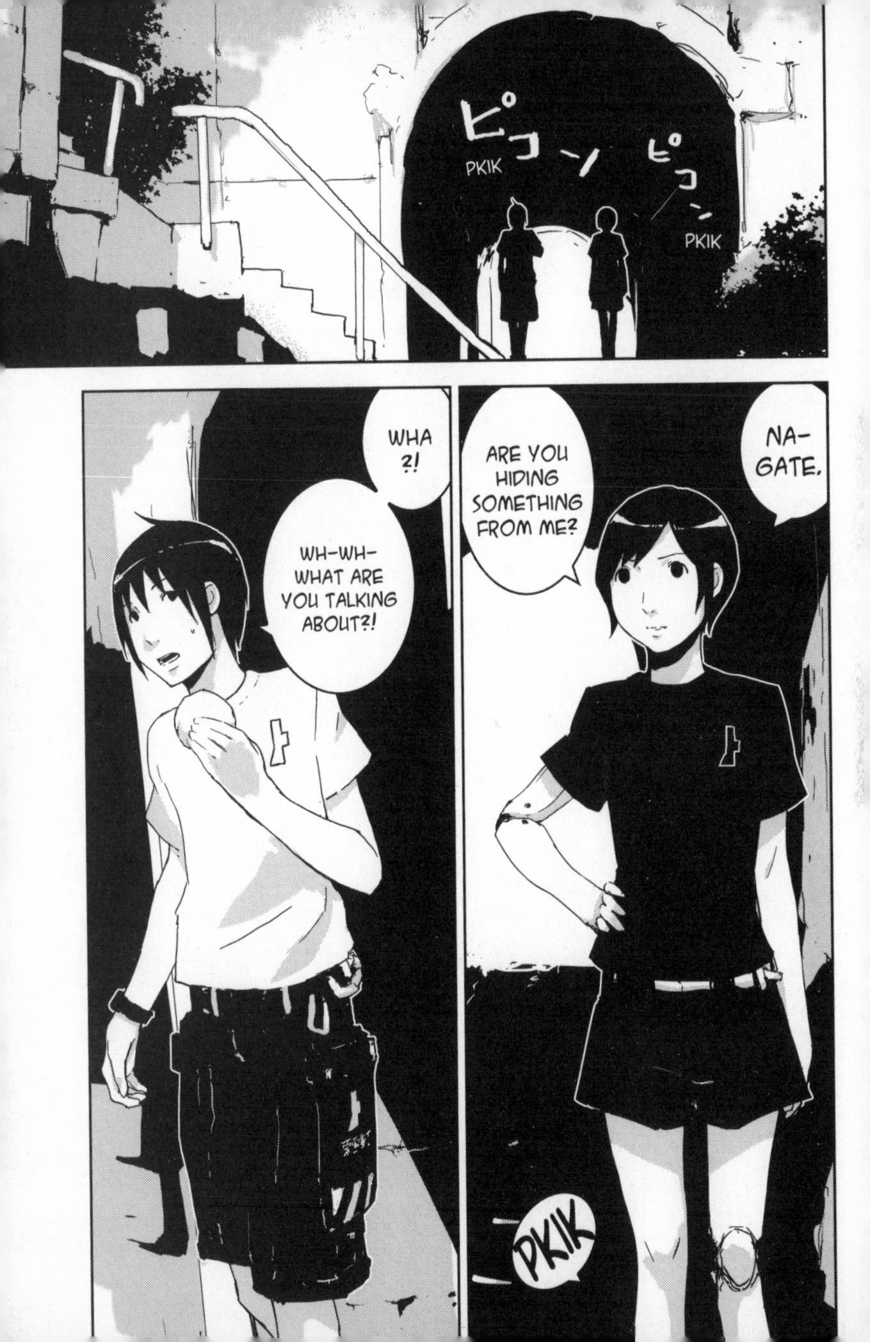
PKIK
PKIK
NA-GATE.
ARE YOU HIDING SOMETHING FROM ME?
PKIK
WHA?!
WH-WH-WHAT ARE YOU TALKING ABOUT?!

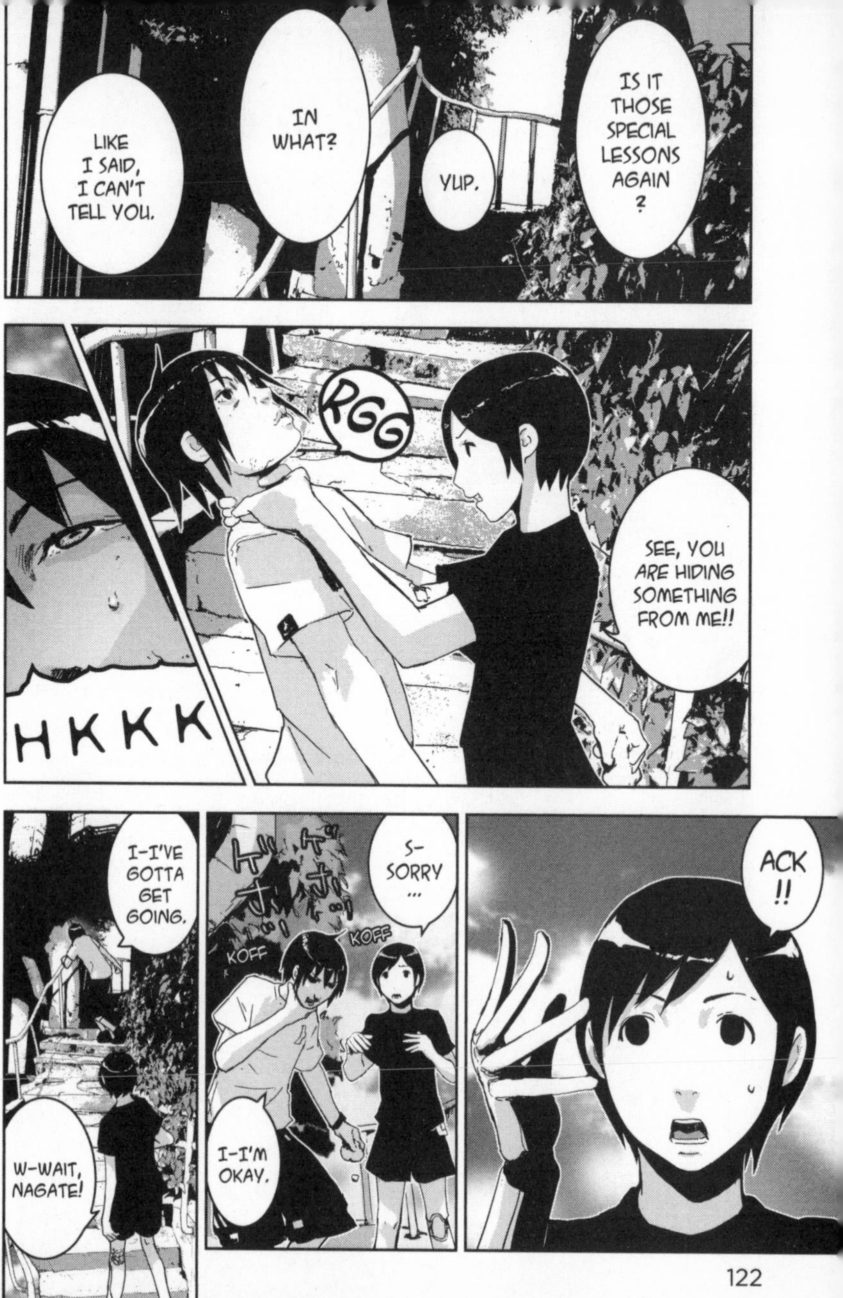
IS IT THOSE SPECIAL LESSONS AGAIN?
YUP.
IN WHAT?
LIKE I SAID, I CAN'T TELL YOU.
SEE, YOU ARE HIDING SOMETHING FROM ME!!
RGG
HKKK
ACK!!
S-SORRY...
KOFF
KOFF
I-I'M OKAY.
I-I'VE GOTTA GET GOING.
W-WAIT, NAGATE!

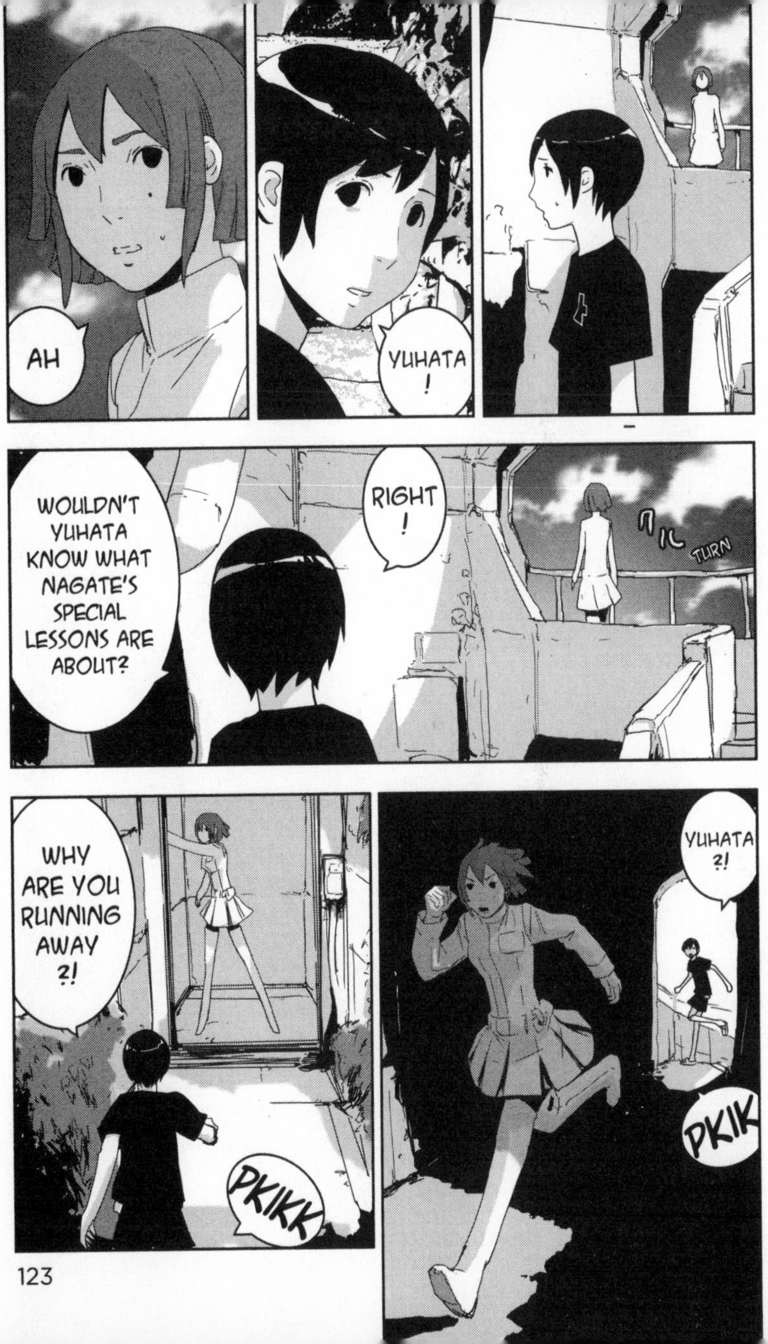
YUHATA !
AH
TURN
RIGHT !
WOULDN'T YUHATA KNOW WHAT NAGATE'S SPECIAL LESSONS ARE ABOUT?
YUHATA ?!
PKIK
WHY ARE YOU RUNNING AWAY ?!
PKIKK

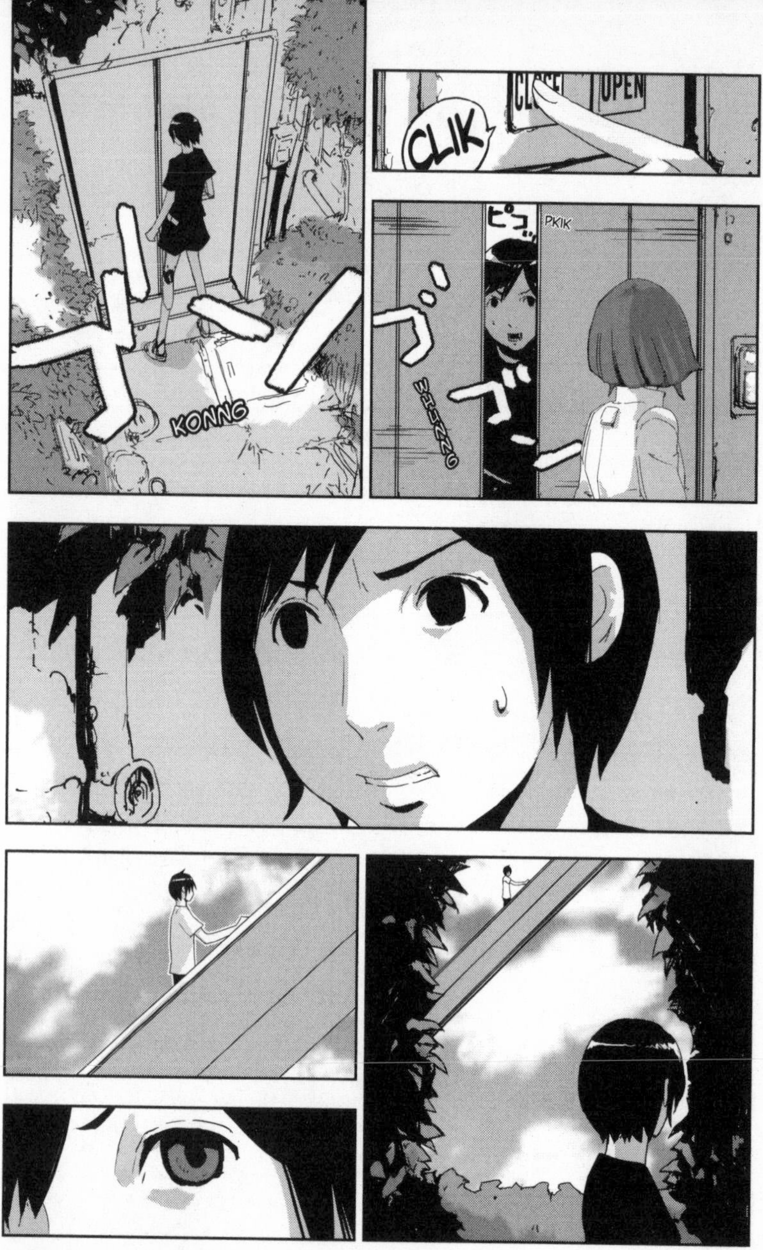
OPEN
CLIK
ピコ
PKIK
WHUNNG
KONNG

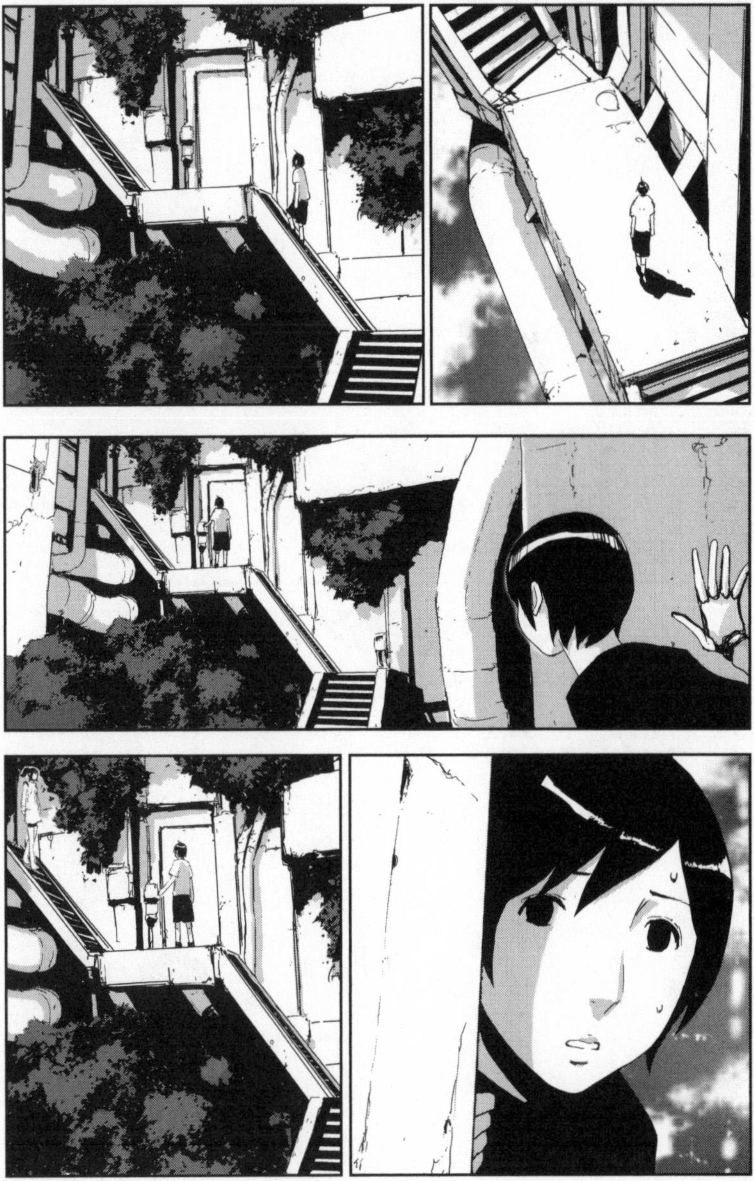

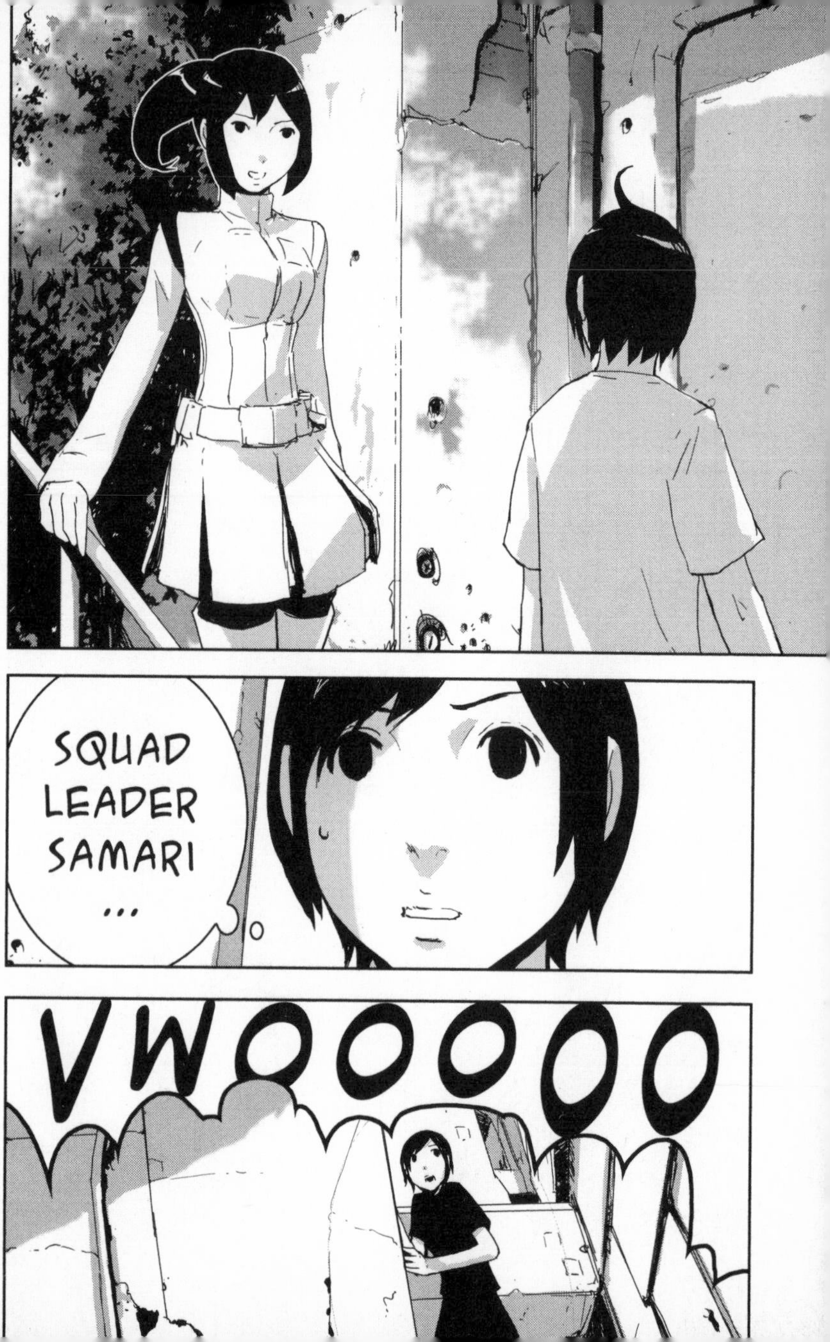
SQUAD LEADER SAMARI ...
VWOOOOO

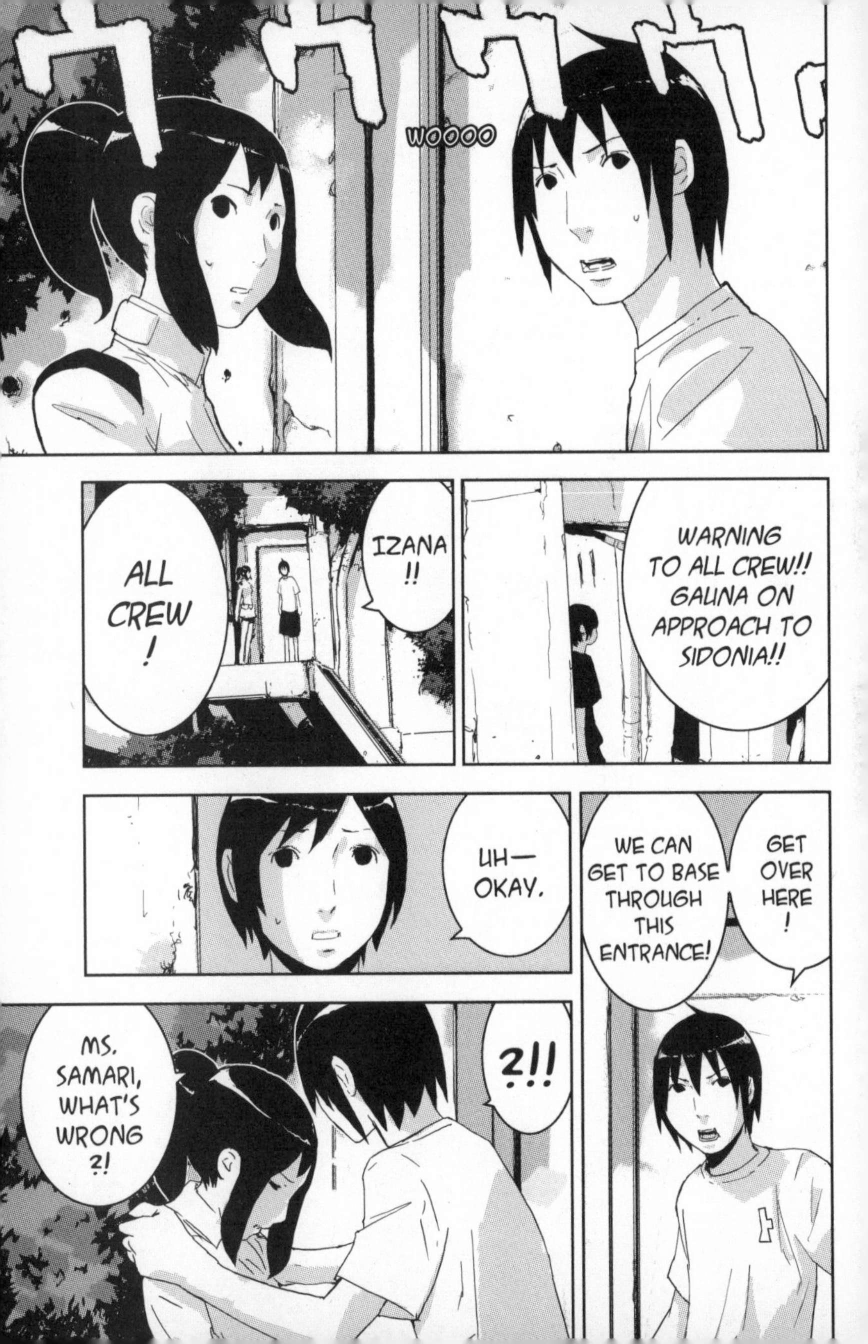
WOOOO
WARNING TO ALL CREW!! GAUNA ON APPROACH TO SIDONIA!!
IZANA!!
ALL CREW!
GET OVER HERE!
WE CAN GET TO BASE THROUGH THIS ENTRANCE!
UH— OKAY.
?!!
MS. SAMARI, WHAT'S WRONG?!

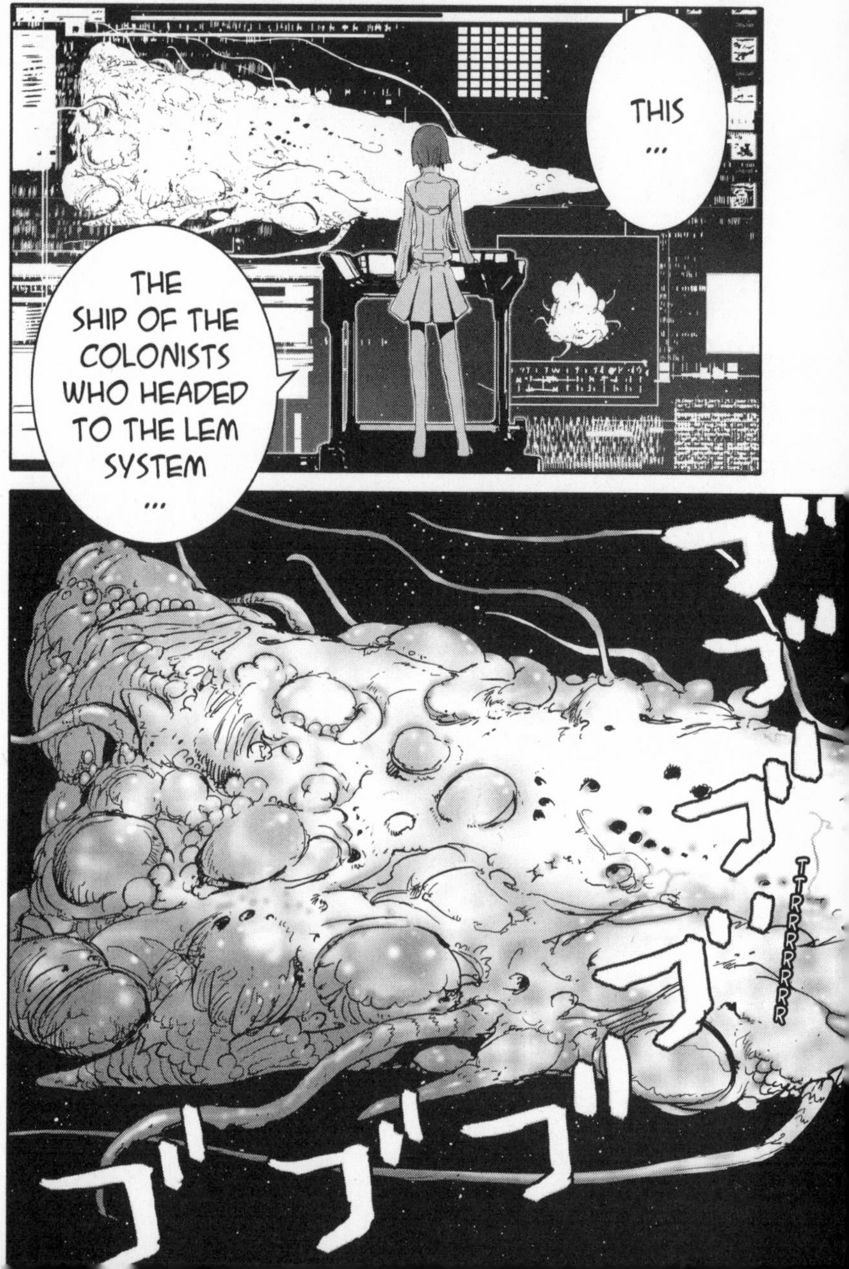
THIS ...
THE SHIP OF THE COLONISTS WHO HEADED TO THE LEM SYSTEM ...
TTRRRRRRR

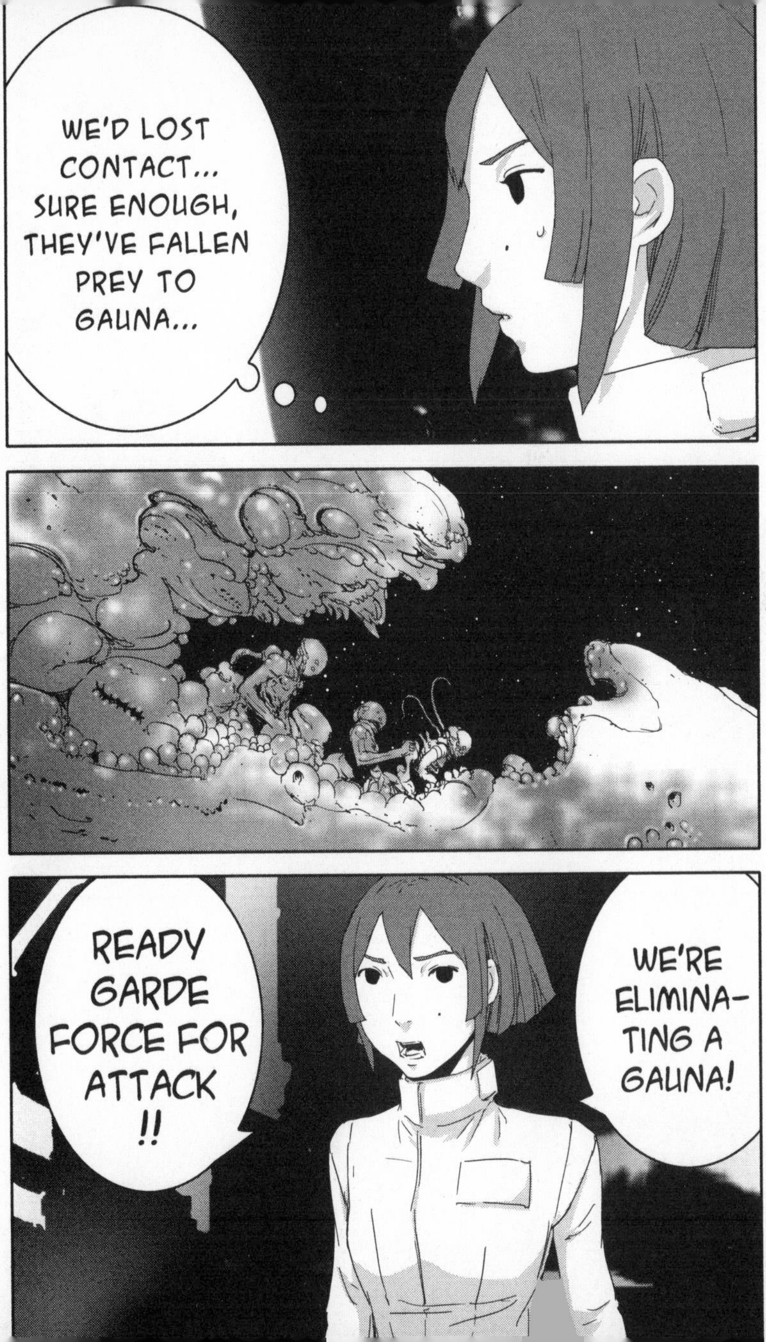
WE'D LOST CONTACT... SURE ENOUGH, THEY'VE FALLEN PREY TO GAUNA...
WE'RE ELIMINA-TING A GAUNA!
READY GARDE FORCE FOR ATTACK !!

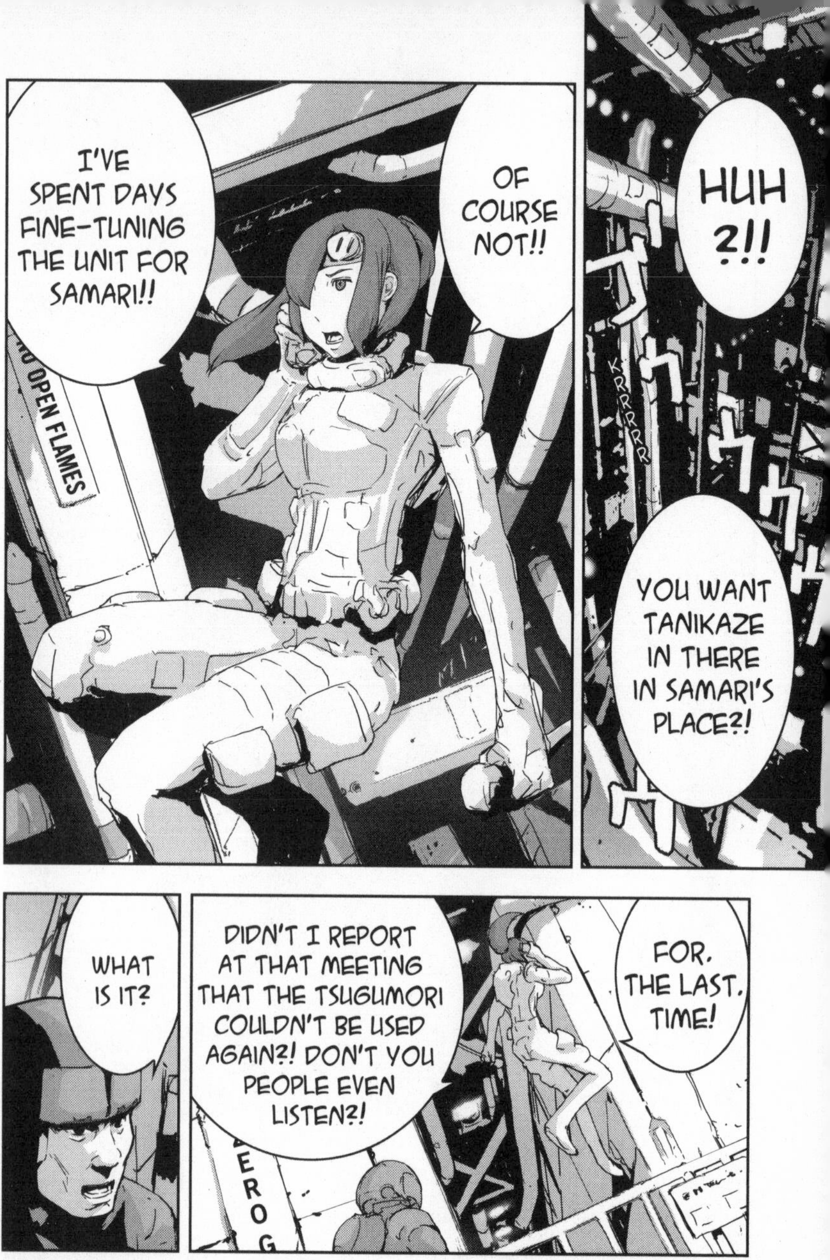
HUH ?!!
KRRRRR
YOU WANT TANIKAZE IN THERE IN SAMARI'S PLACE?!
OF COURSE NOT!!
I'VE SPENT DAYS FINE-TUNING THE UNIT FOR SAMARI!!
OPEN FLAMES
FOR. THE LAST. TIME!
DIDN'T I REPORT AT THAT MEETING THAT THE TSUGUMORI COULDN'T BE USED AGAIN?! DON'T YOU PEOPLE EVEN LISTEN?!
WHAT IS IT?

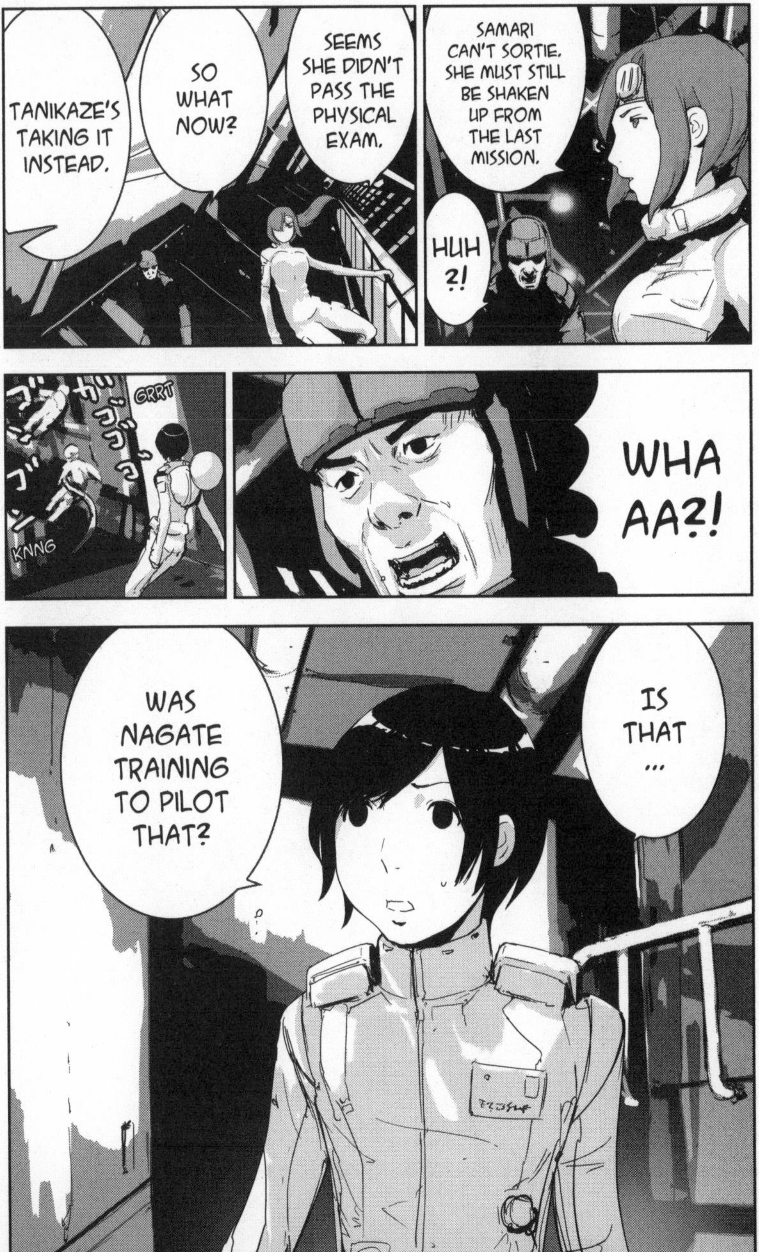
SAMARI CAN'T SORTIE. SHE MUST STILL BE SHAKEN UP FROM THE LAST MISSION.
HUH ?!
SEEMS SHE DIDN'T PASS THE PHYSICAL EXAM.
SO WHAT NOW?
TANIKAZE'S TAKING IT INSTEAD.
WHA AA?!
GRRT
KNNG
IS THAT ...
WAS NAGATE TRAINING TO PILOT THAT?

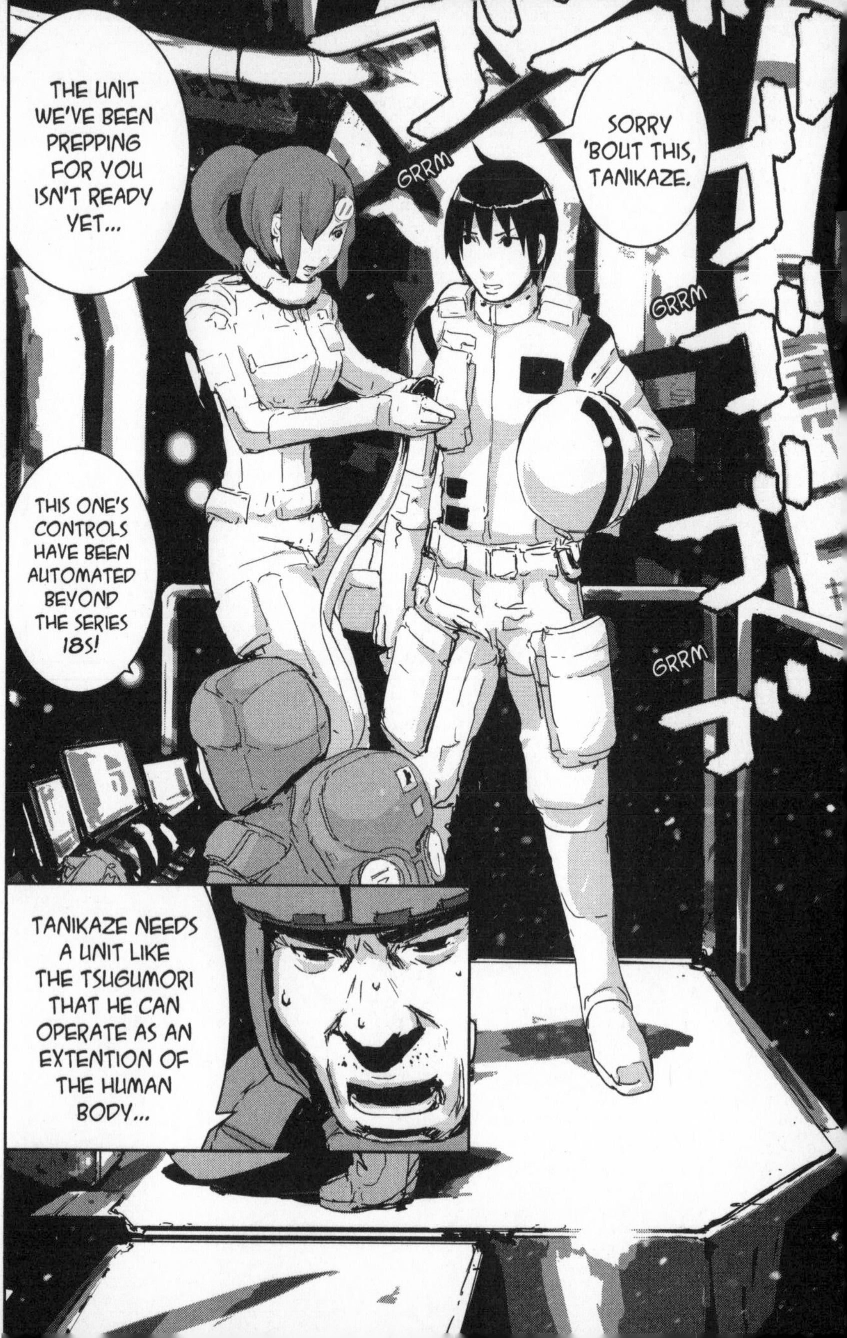
SORRY 'BOUT THIS, TANIKAZE.
GRRM
THE UNIT WE'VE BEEN PREPPING FOR YOU ISN'T READY YET...
GRRM
GRRM
THIS ONE'S CONTROLS HAVE BEEN AUTOMATED BEYOND THE SERIES 18S!
TANIKAZE NEEDS A UNIT LIKE THE TSUGUMORI THAT HE CAN OPERATE AS AN EXTENTION OF THE HUMAN BODY...

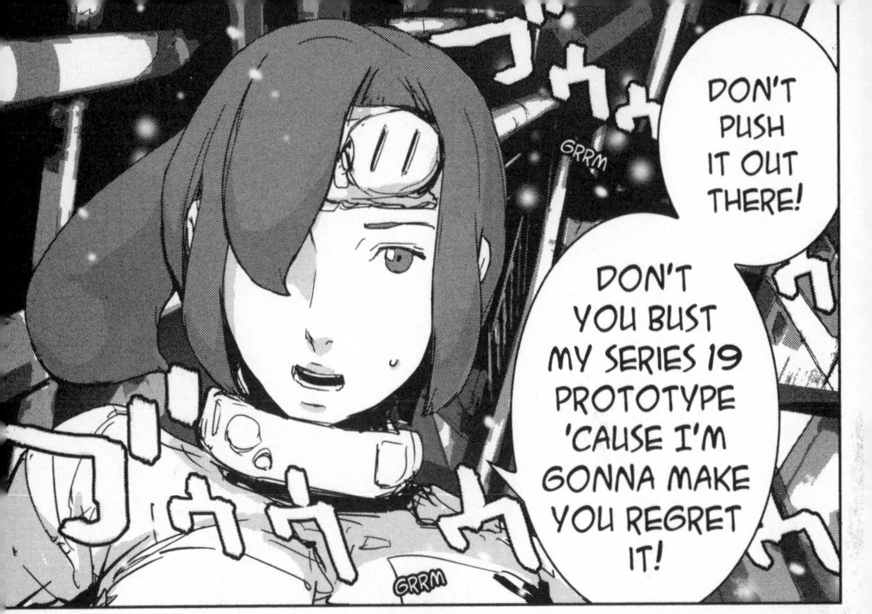
DON'T PUSH IT OUT THERE!
GRRM
DON'T YOU BUST MY SERIES 19 PROTOTYPE 'CAUSE I'M GONNA MAKE YOU REGRET IT!
GRRM

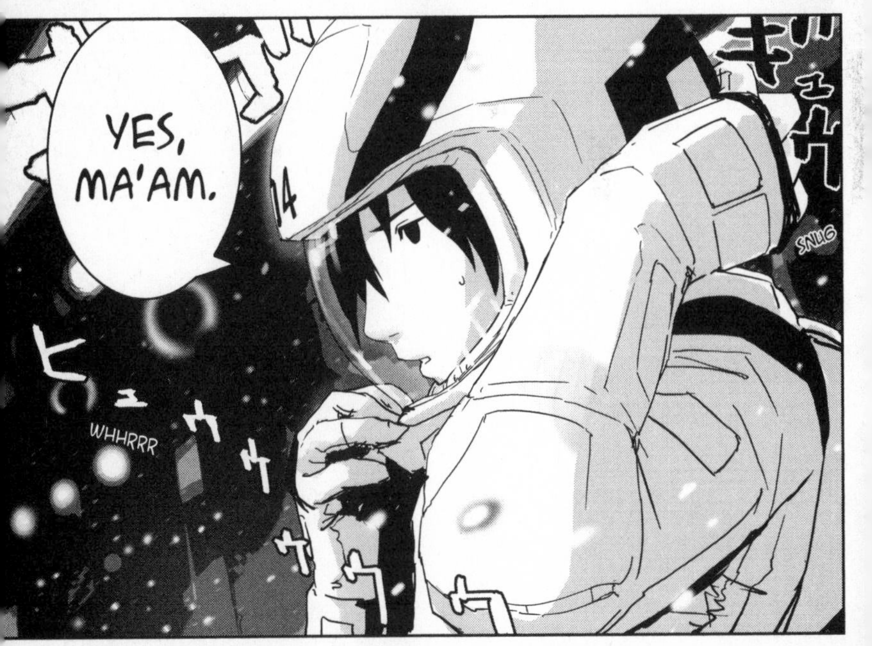
SNUG
YES, MA'AM.
WHHRRR

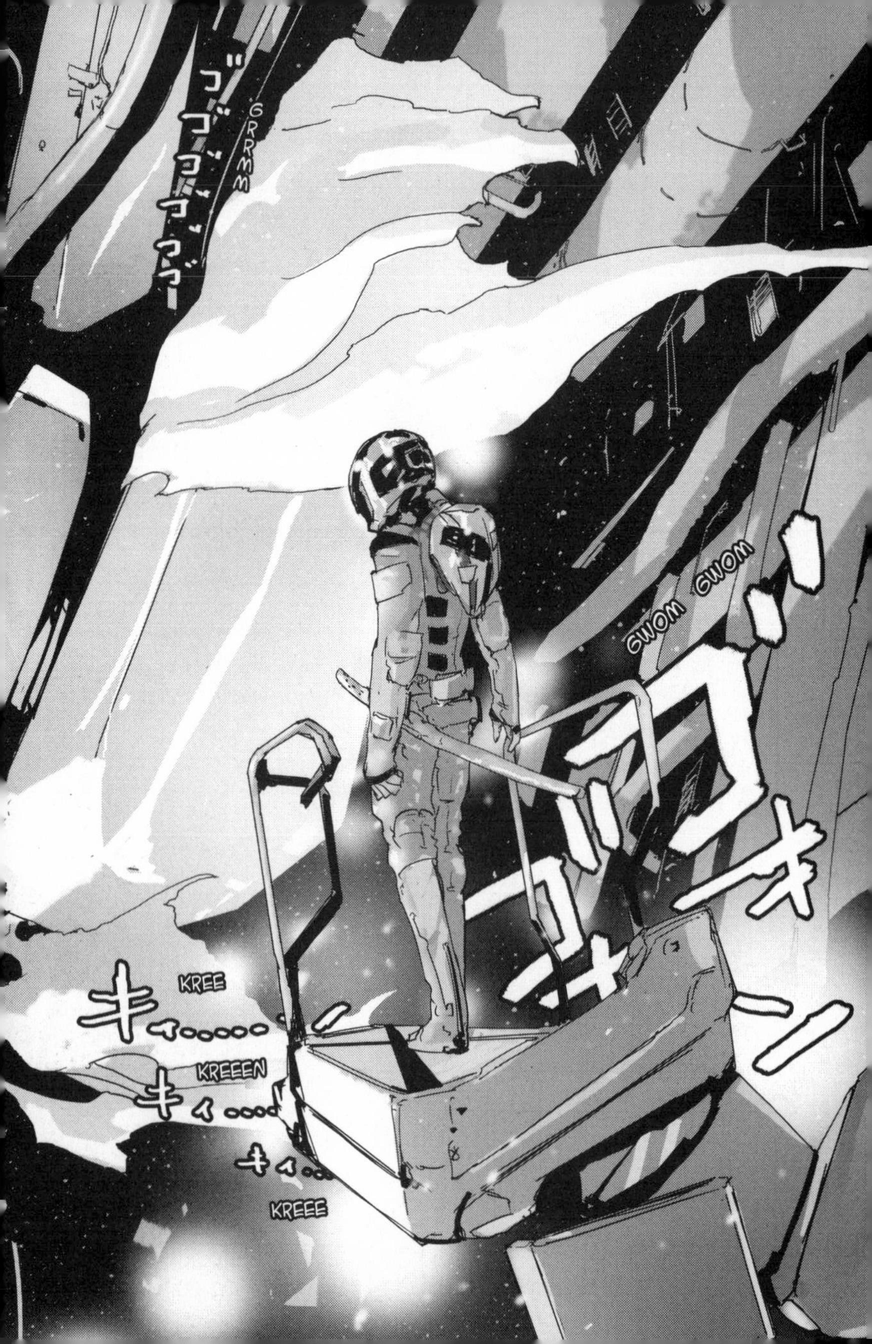
GRRMM
GWOM GWOM
KREE
キィ……
KREEEN
キィ……
キィ…
KREEE

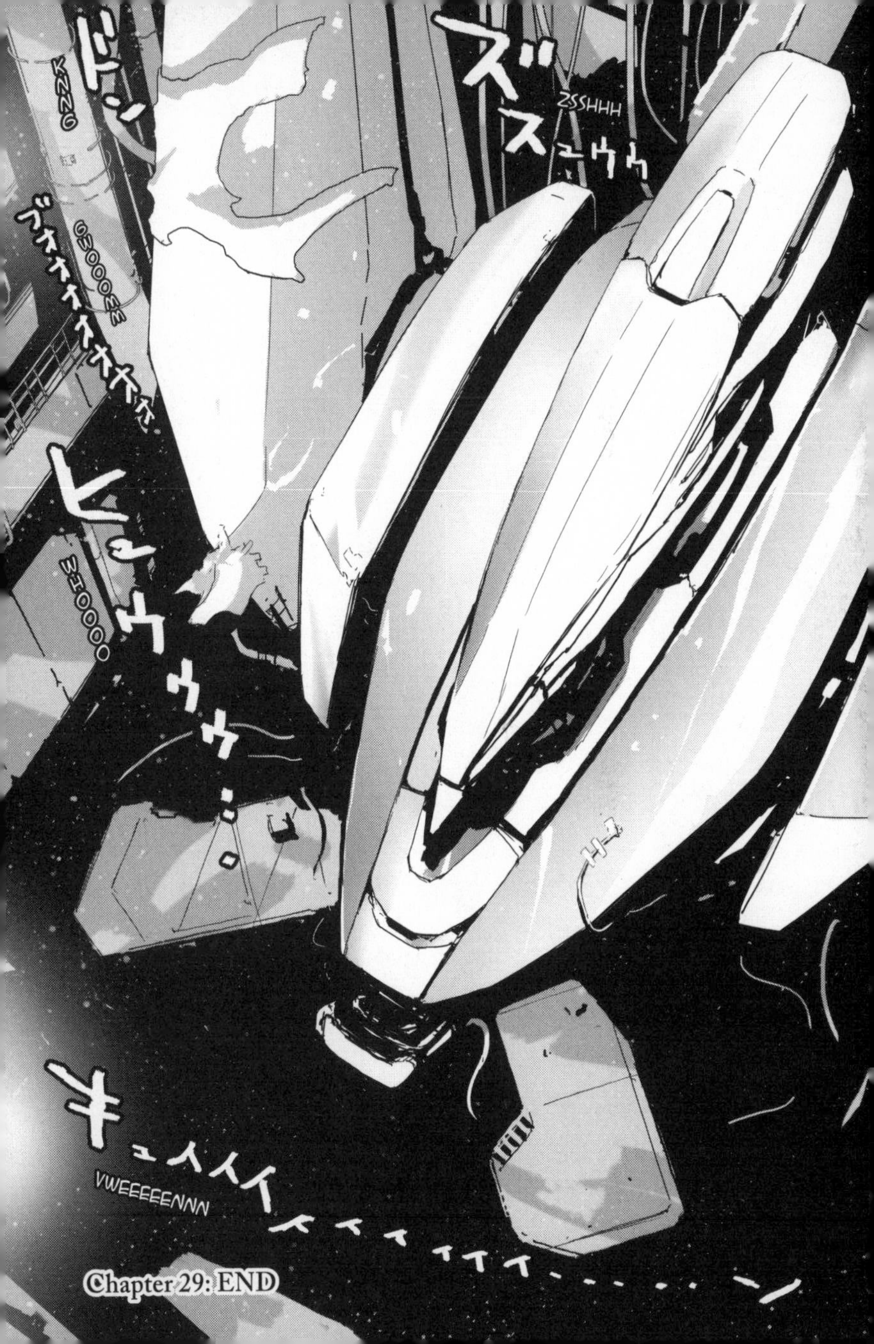
ズスュウウ
ZSSHHH
ドン
KNNG
ブオオオオオオオ
GVOOOOMM
ヒュウウウ
WHOOOO
キュイイイイイイイイイ――――ン
VWEEEEEENNN

Chapter 29: END

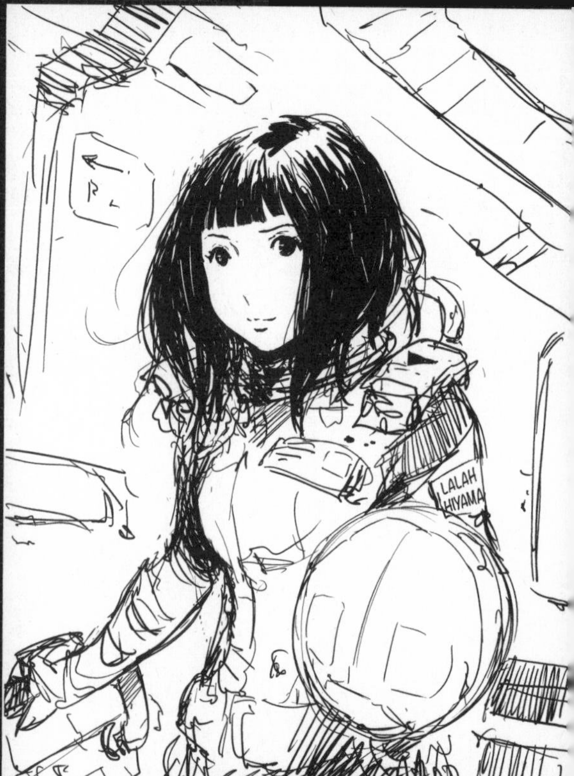
LALAH
HIYAMA

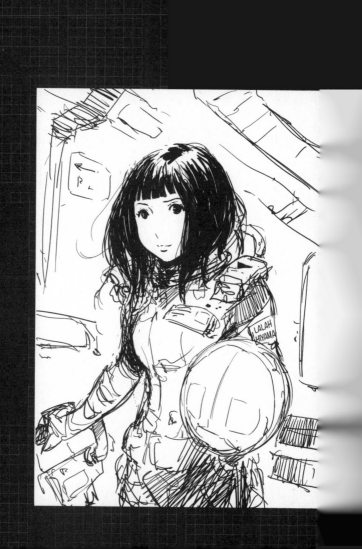

Chapter 30: Ren Honoka's Resolution

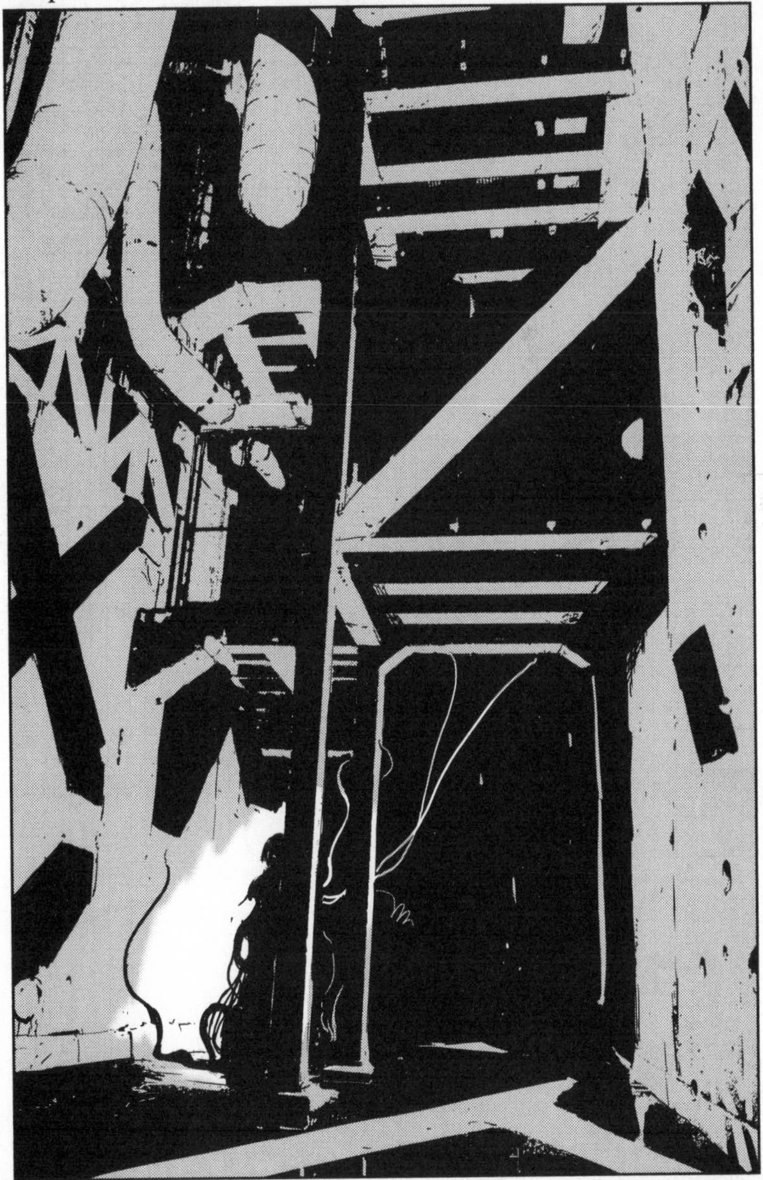

One Hundred Sights of Sidonia Part Twenty-Four:
Former Research Facility

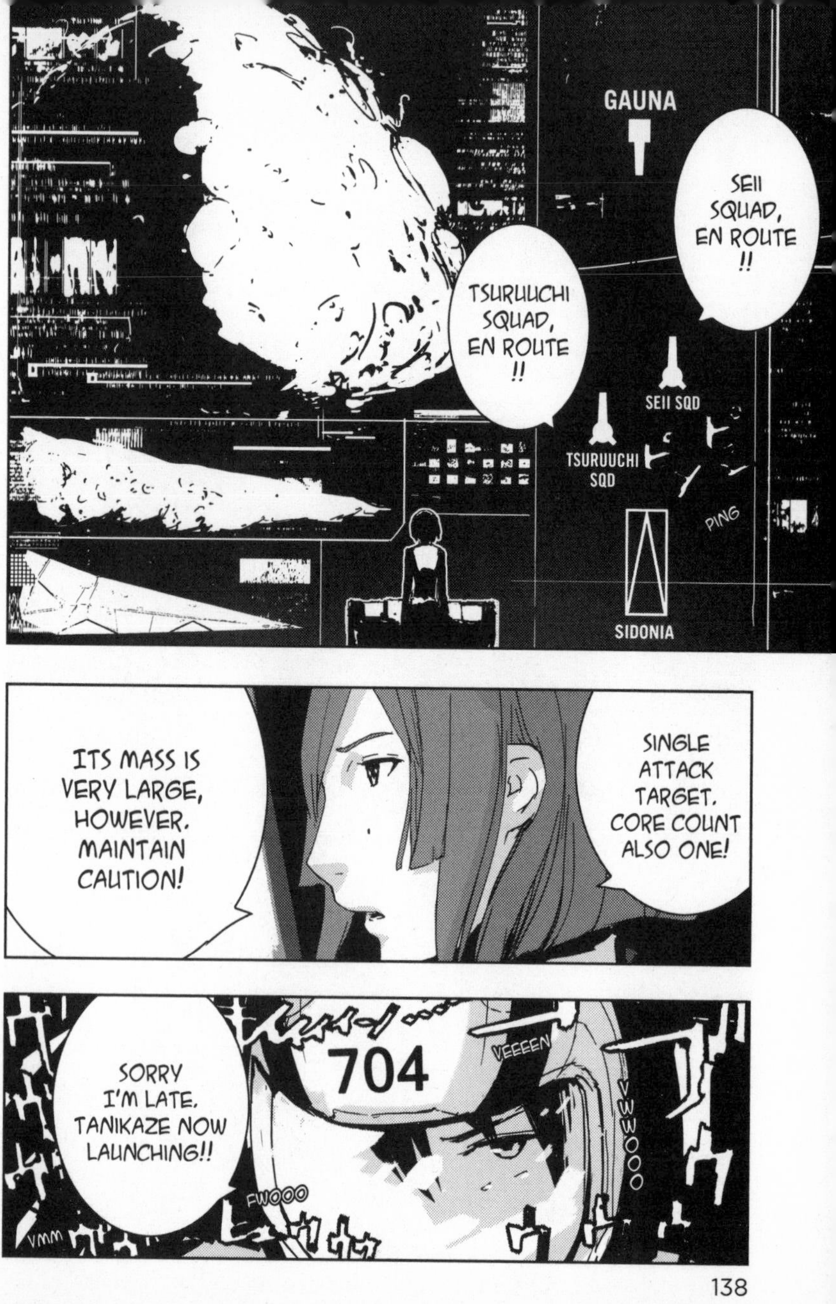
GAUNA
SEII SQUAD, EN ROUTE !!
TSURUUCHI SQUAD, EN ROUTE !!
SEII SQD
TSURUUCHI SQD
PING
SIDONIA
SINGLE ATTACK TARGET. CORE COUNT ALSO ONE!
ITS MASS IS VERY LARGE, HOWEVER. MAINTAIN CAUTION!
704
VEEEEN
VWWOOO
SORRY I'M LATE. TANIKAZE NOW LAUNCHING!!
FWOOO
VMM

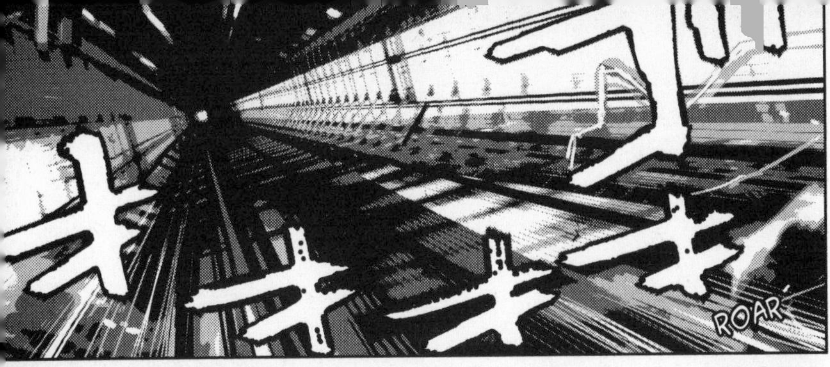
ROAR
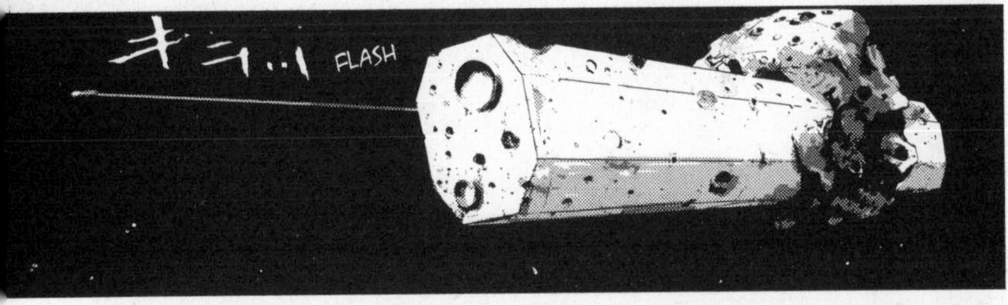
FLASH
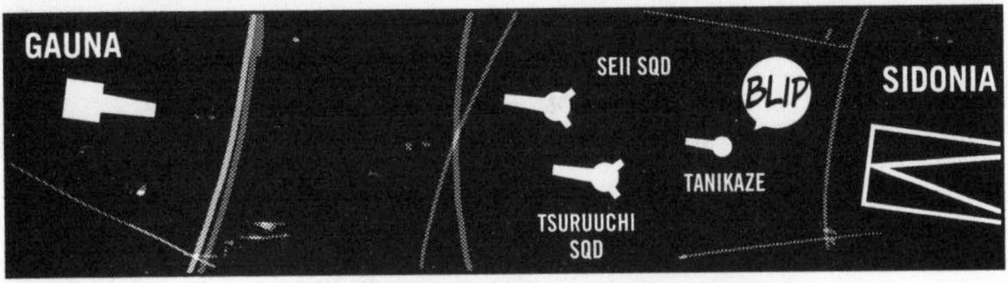
SIDONIA
BLIP
SEII SQD
TANIKAZE
TSURUUCHI SQD
GAUNA
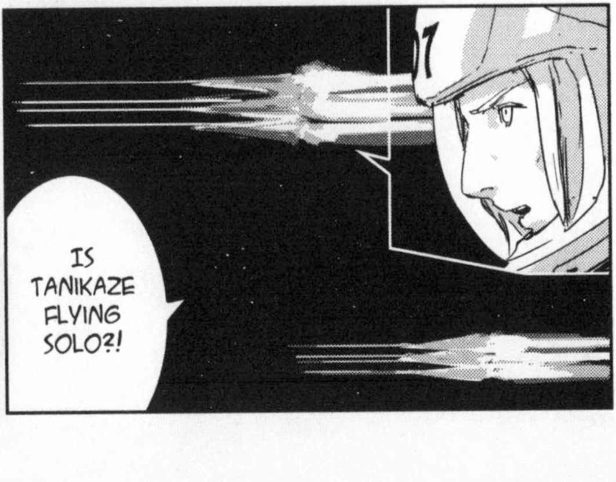
IS TANIKAZE FLYING SOLO?!
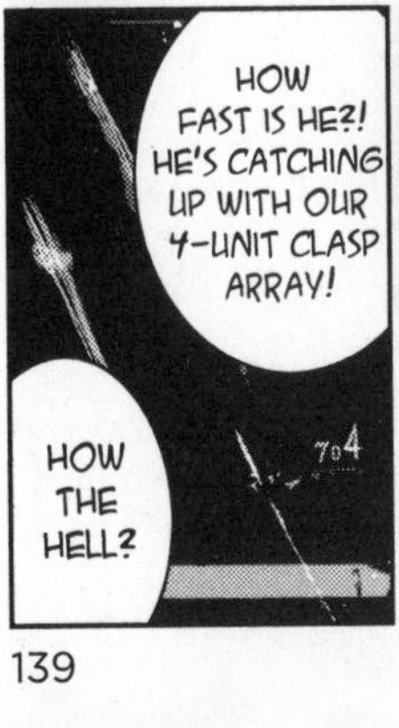
HOW FAST IS HE?! HE'S CATCHING UP WITH OUR 4-UNIT CLASP ARRAY!
HOW THE HELL?

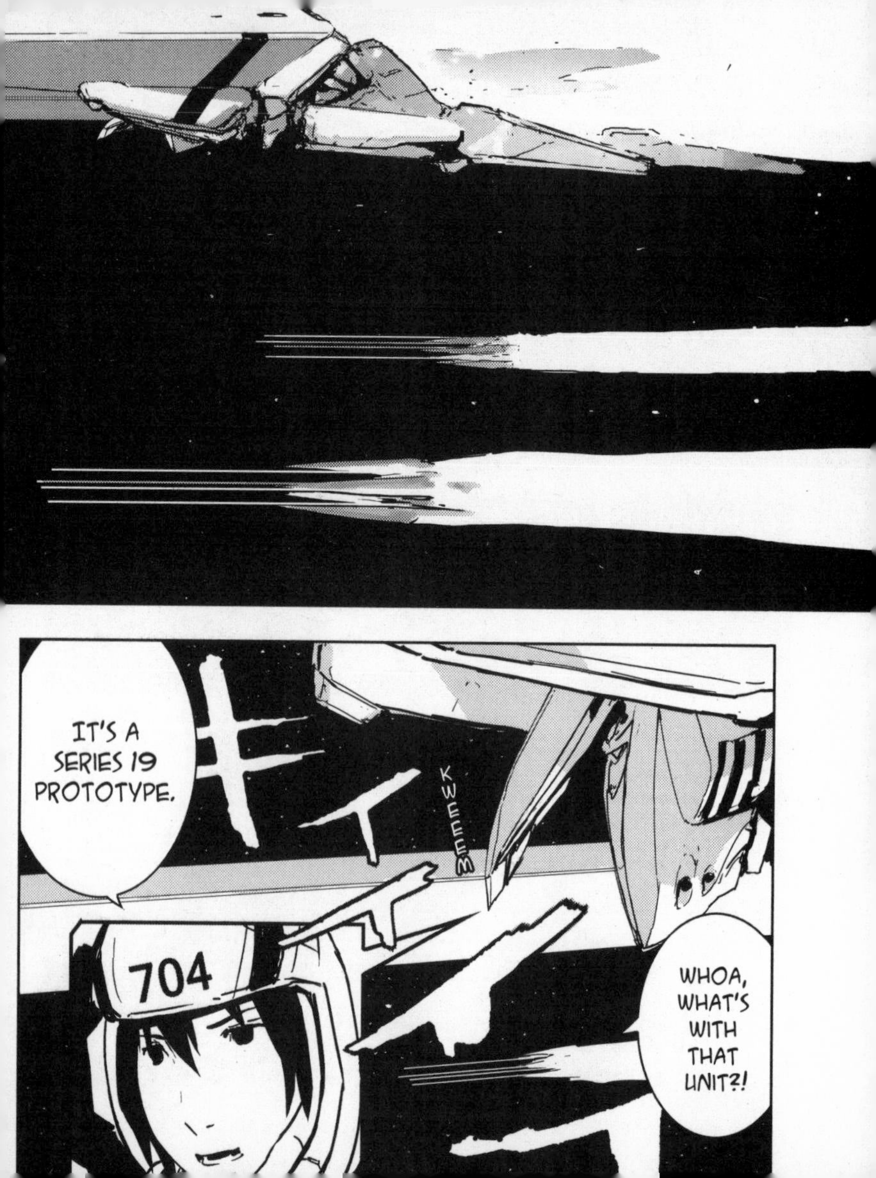
KWEEEM
WHOA, WHAT'S WITH THAT UNIT?!
IT'S A SERIES 19 PROTOTYPE.
704

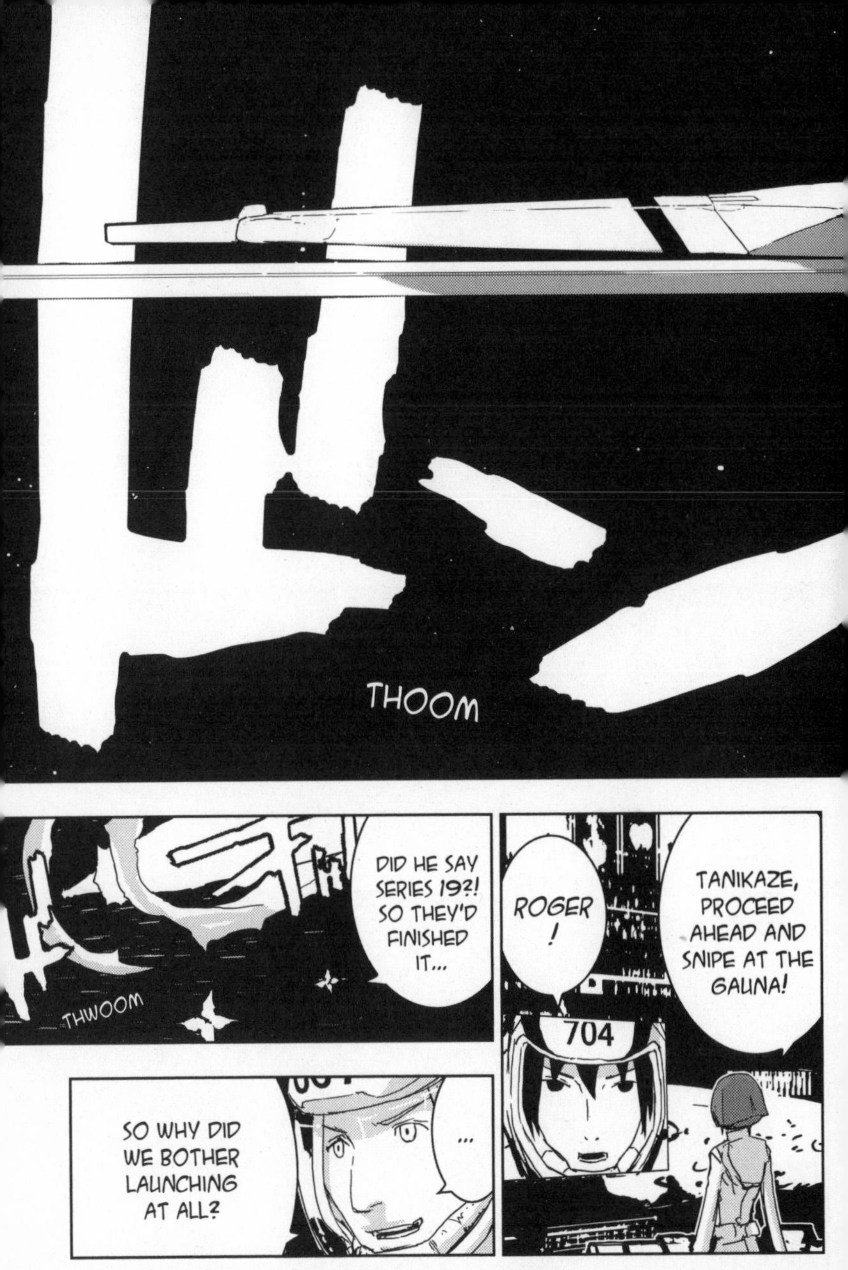
THOOM
TANIKAZE, PROCEED AHEAD AND SNIPE AT THE GAUNA!
ROGER!
704
DID HE SAY SERIES 19?! SO THEY'D FINISHED IT...
THWOOM
...
SO WHY DID WE BOTHER LAUNCHING AT ALL?

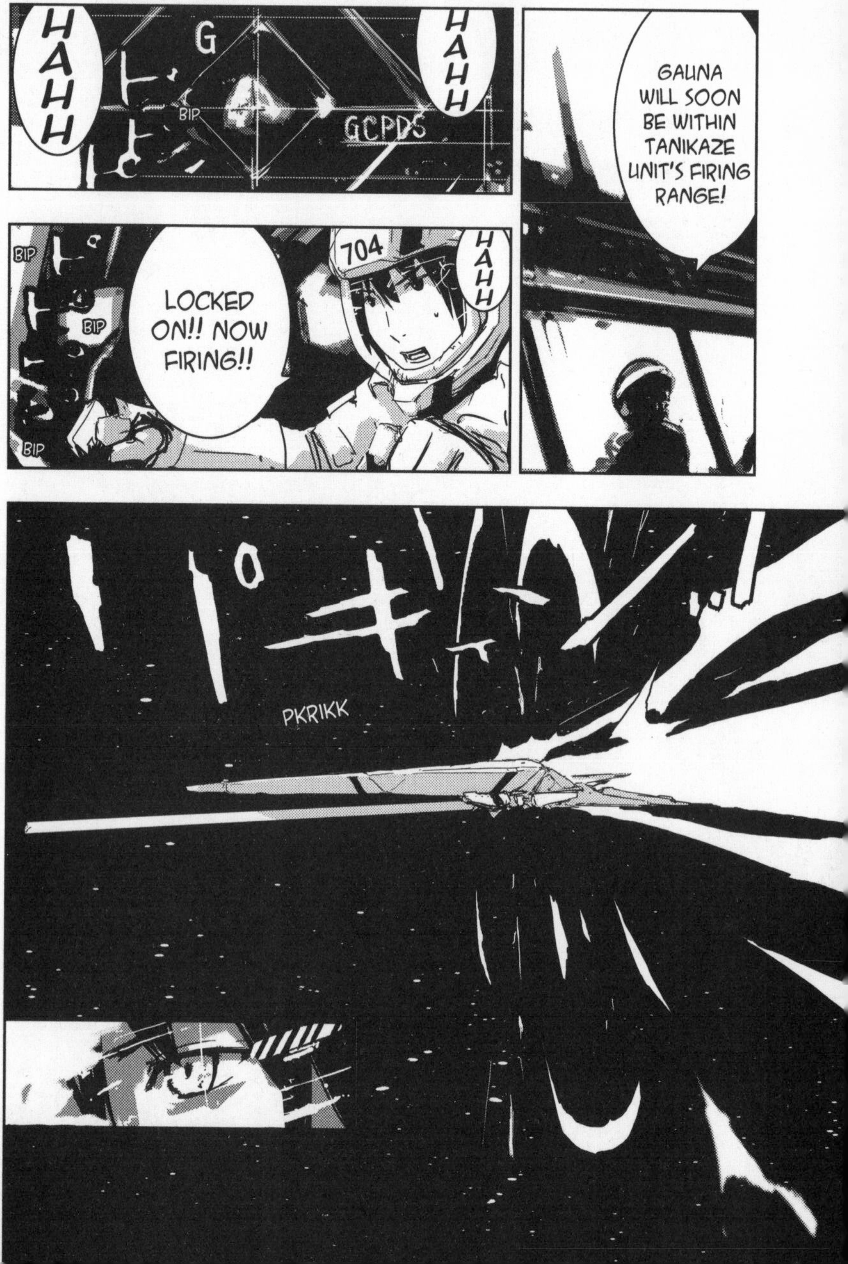
GAUNA WILL SOON BE WITHIN TANIKAZE UNIT'S FIRING RANGE!
HAHH
G
HAHH
GCPDS
BIP
HAHH
704
LOCKED ON!! NOW FIRING!!
BIP
BIP
BIP
PKRIKK

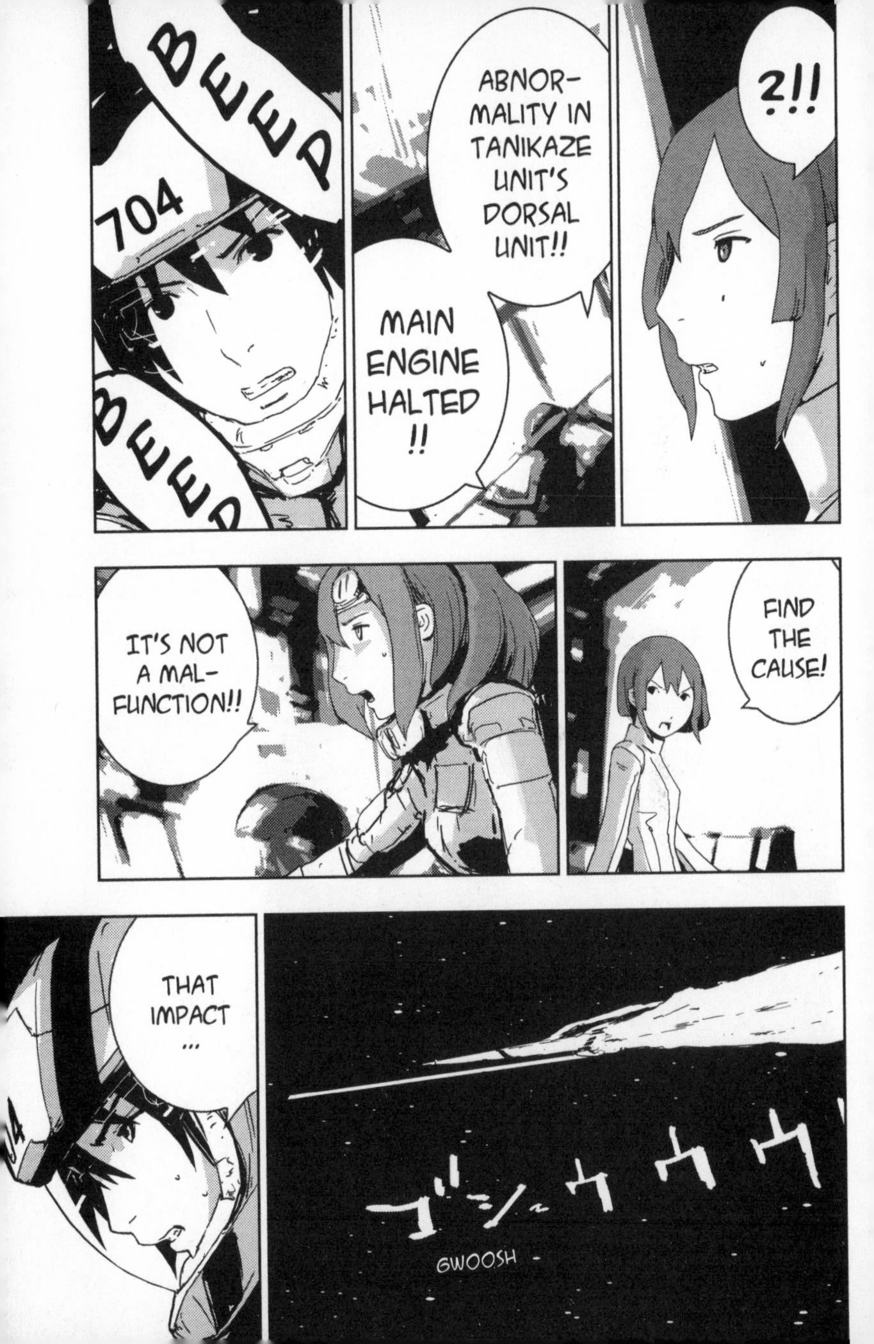
?!!
ABNOR-MALITY IN TANIKAZE UNIT'S DORSAL UNIT!!
MAIN ENGINE HALTED !!
BEEP
704
BEEP
FIND THE CAUSE!
IT'S NOT A MAL-FUNCTION!!
ゴシュウウウ
GWOOSH
THAT IMPACT ...

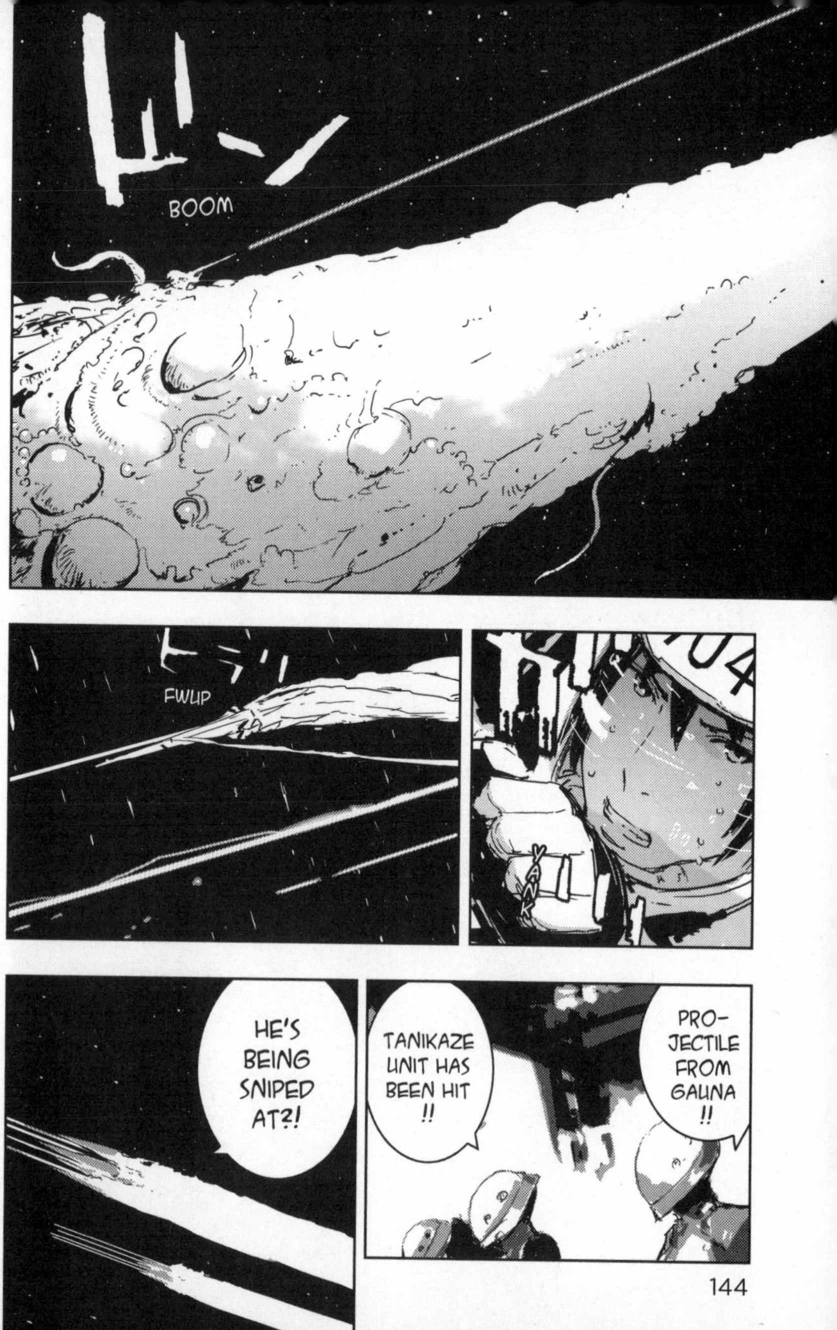
BOOM
YANK
FWUP
PRO-JECTILE FROM GAUNA!!
TANIKAZE UNIT HAS BEEN HIT!!
HE'S BEING SNIPED AT?!

A GAUNA, FIRING LONG-RANGE SHOTS...

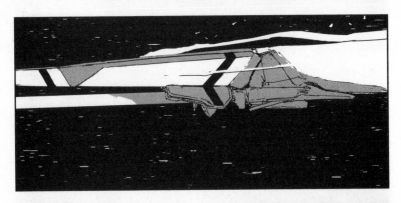

CHIK

704

ピ
ピ
ピ
ピ
ト

BIP

BIP

BIP

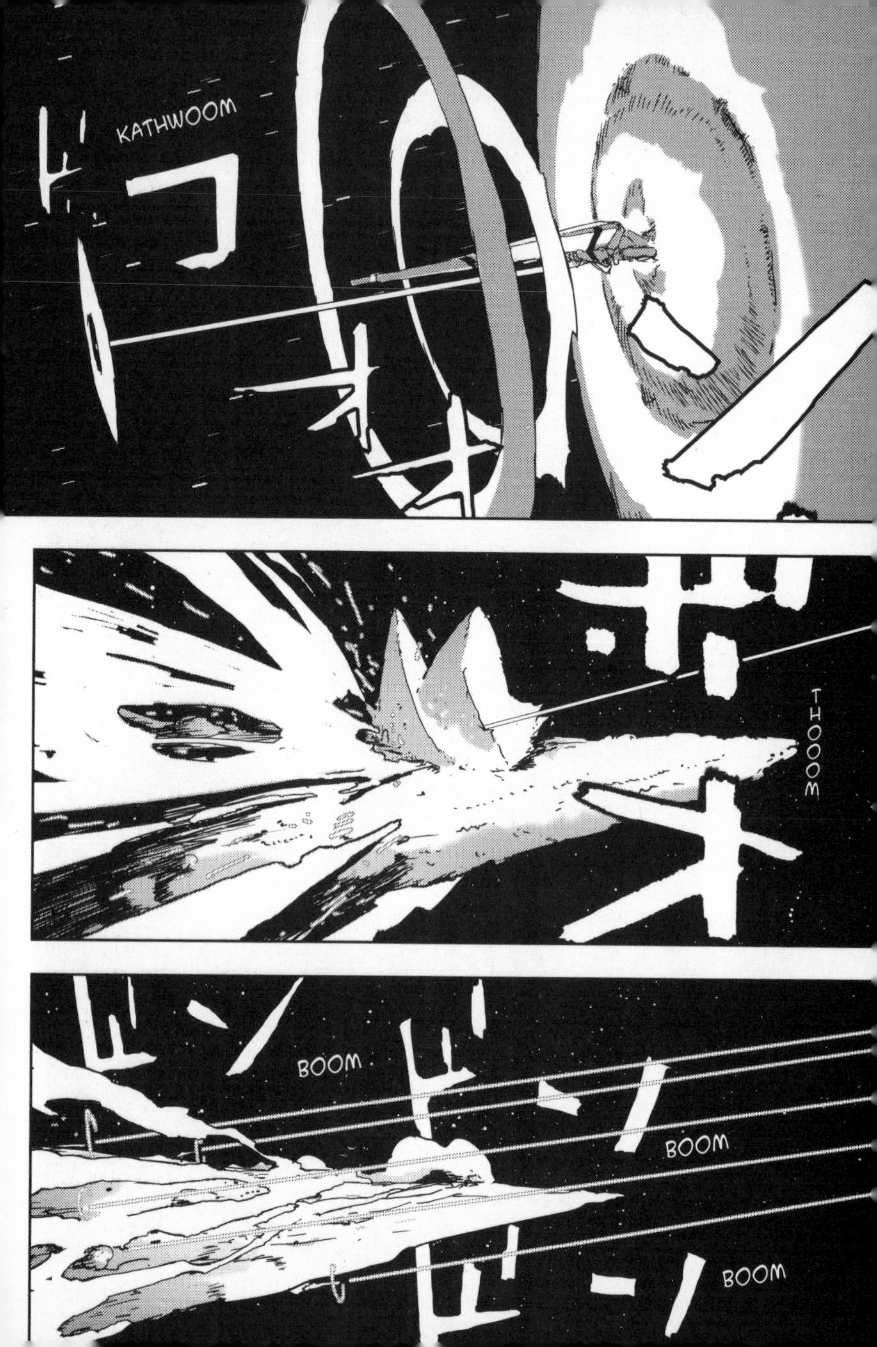
KATHWOOM
THOOOM
BOOM
BOOM
BOOM

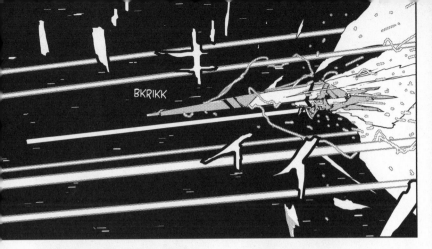

BKRIKK
DIRECT HIT ON GAUNA WITH GCPDS!!
TANIKAZE UNIT HAS ALSO BEEN HIT!!
HOWEVER, DAMAGE TO PROTOTYPE IS MINIMAL!
THE GAUNA CORE HASN'T BEEN DESTROYED?!!
THE GCPDS ROUND WAS STOPPED AT THE GAUNA'S SURFACE!!

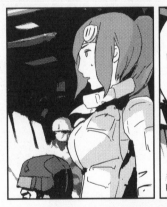

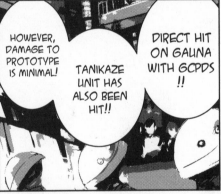

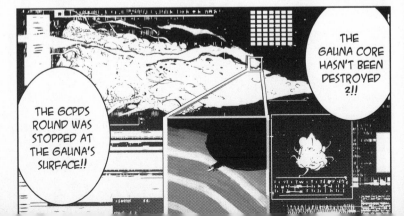

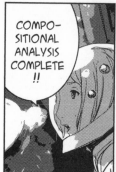

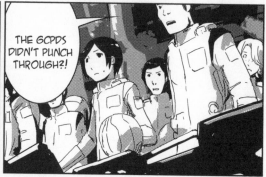

THE GCPDS DIDN'T PUNCH THROUGH?!
COMPO-SITIONAL ANALYSIS COMPLETE !!
FWAP
WHAT THE...
PLACENTA
SHELL
GCPDS
ORDINARY PLACENTA AND PLACENTA TRANSFORMED INTO PSEUDO-CORE SHELLS ALTERNATING IN LAYERS...
WHICH MEANS ANY ROUNDS THAT MANAGE TO PIERCE THROUGH THE FIRST SHELL SIMPLY GET STOPPED AT THE SECOND PLACENTA...

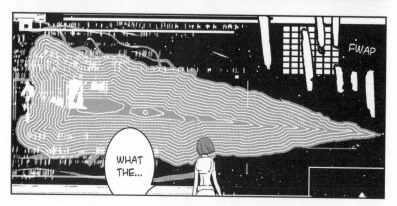

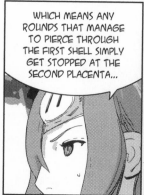

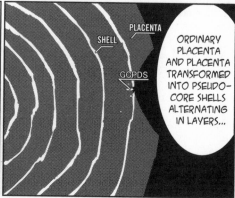

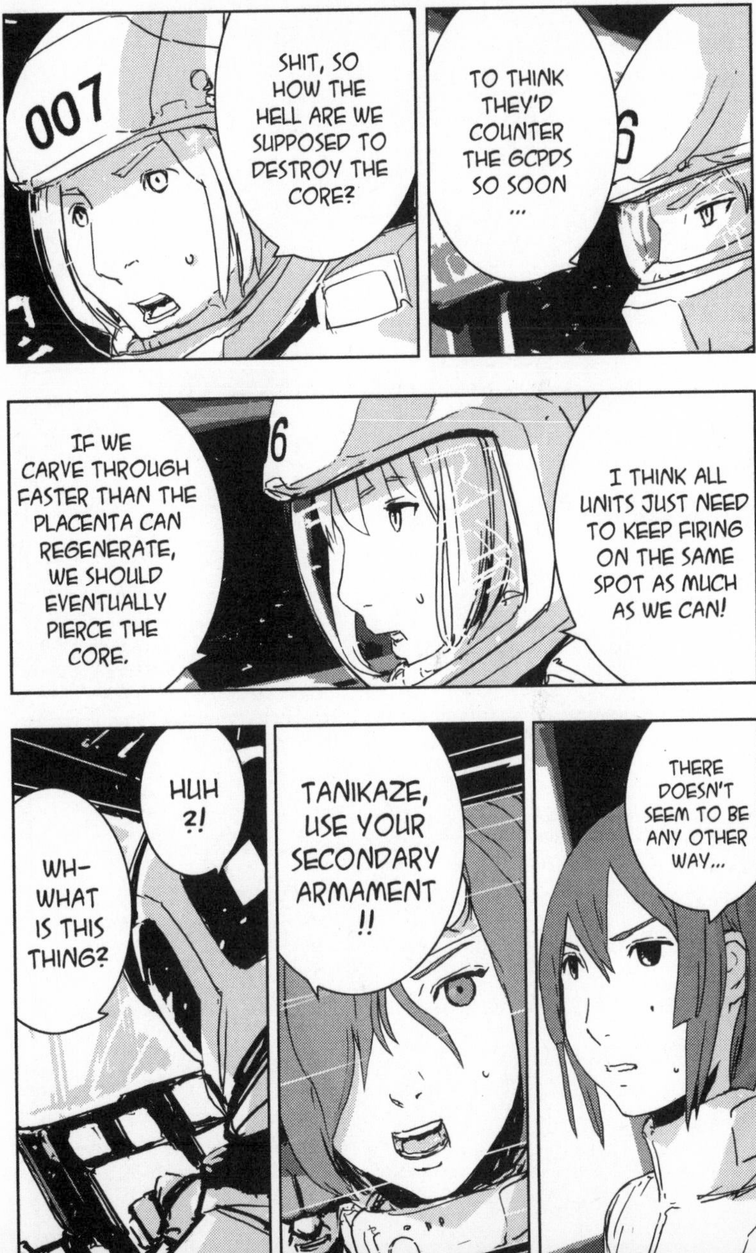
TO THINK THEY'D COUNTER THE GCPDS SO SOON ...
SHIT, SO HOW THE HELL ARE WE SUPPOSED TO DESTROY THE CORE?
007
I THINK ALL UNITS JUST NEED TO KEEP FIRING ON THE SAME SPOT AS MUCH AS WE CAN!
IF WE CARVE THROUGH FASTER THAN THE PLACENTA CAN REGENERATE, WE SHOULD EVENTUALLY PIERCE THE CORE.
THERE DOESN'T SEEM TO BE ANY OTHER WAY...
TANIKAZE, USE YOUR SECONDARY ARMAMENT !!
HUH ?!
WH- WHAT IS THIS THING?

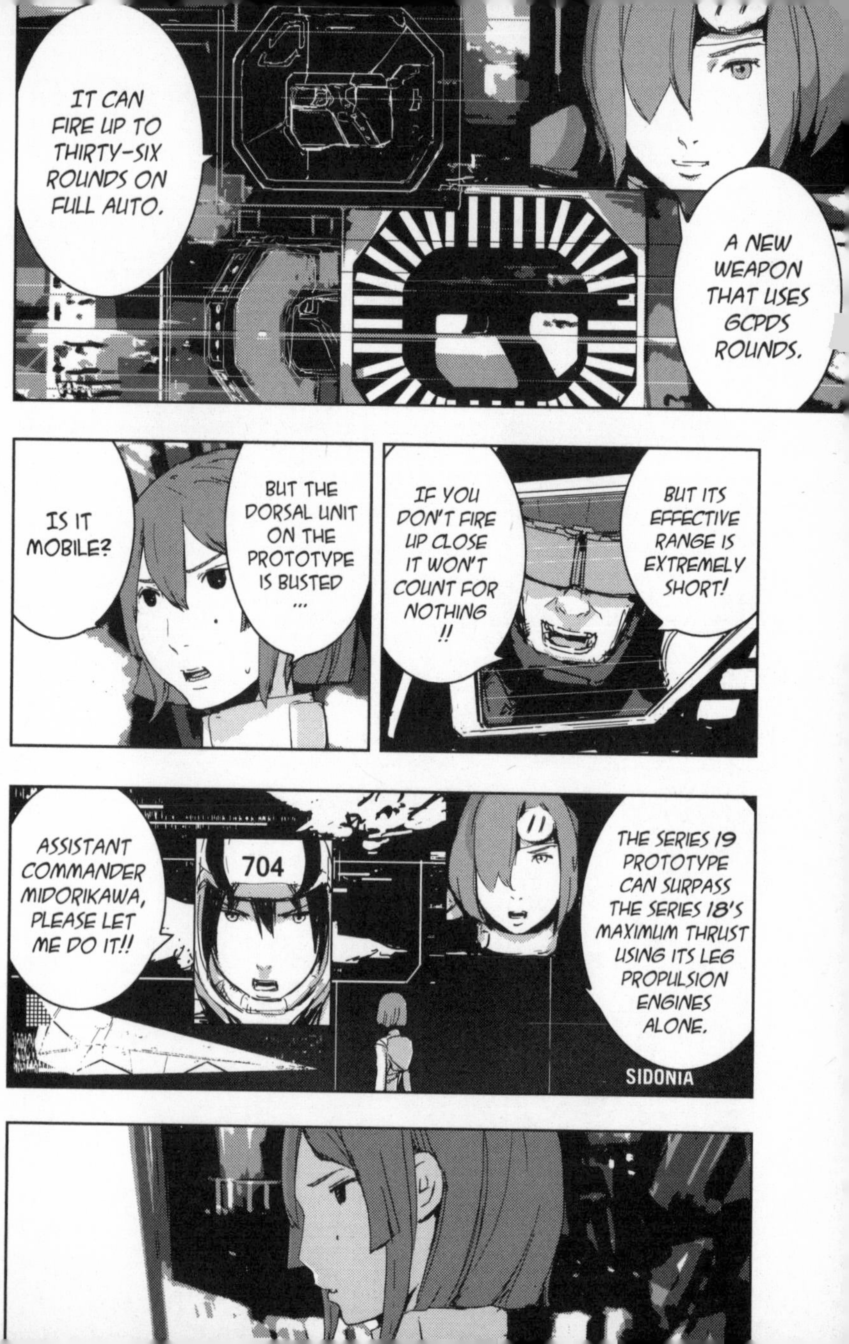
A NEW WEAPON THAT USES GCPDS ROUNDS.
IT CAN FIRE UP TO THIRTY-SIX ROUNDS ON FULL AUTO.
BUT ITS EFFECTIVE RANGE IS EXTREMELY SHORT!
IF YOU DON'T FIRE UP CLOSE IT WON'T COUNT FOR NOTHING!!
BUT THE DORSAL UNIT ON THE PROTOTYPE IS BUSTED...
IS IT MOBILE?
THE SERIES 19 PROTOTYPE CAN SURPASS THE SERIES 18'S MAXIMUM THRUST USING ITS LEG PROPULSION ENGINES ALONE.
SIDONIA
704
ASSISTANT COMMANDER MIDORIKAWA, PLEASE LET ME DO IT!!

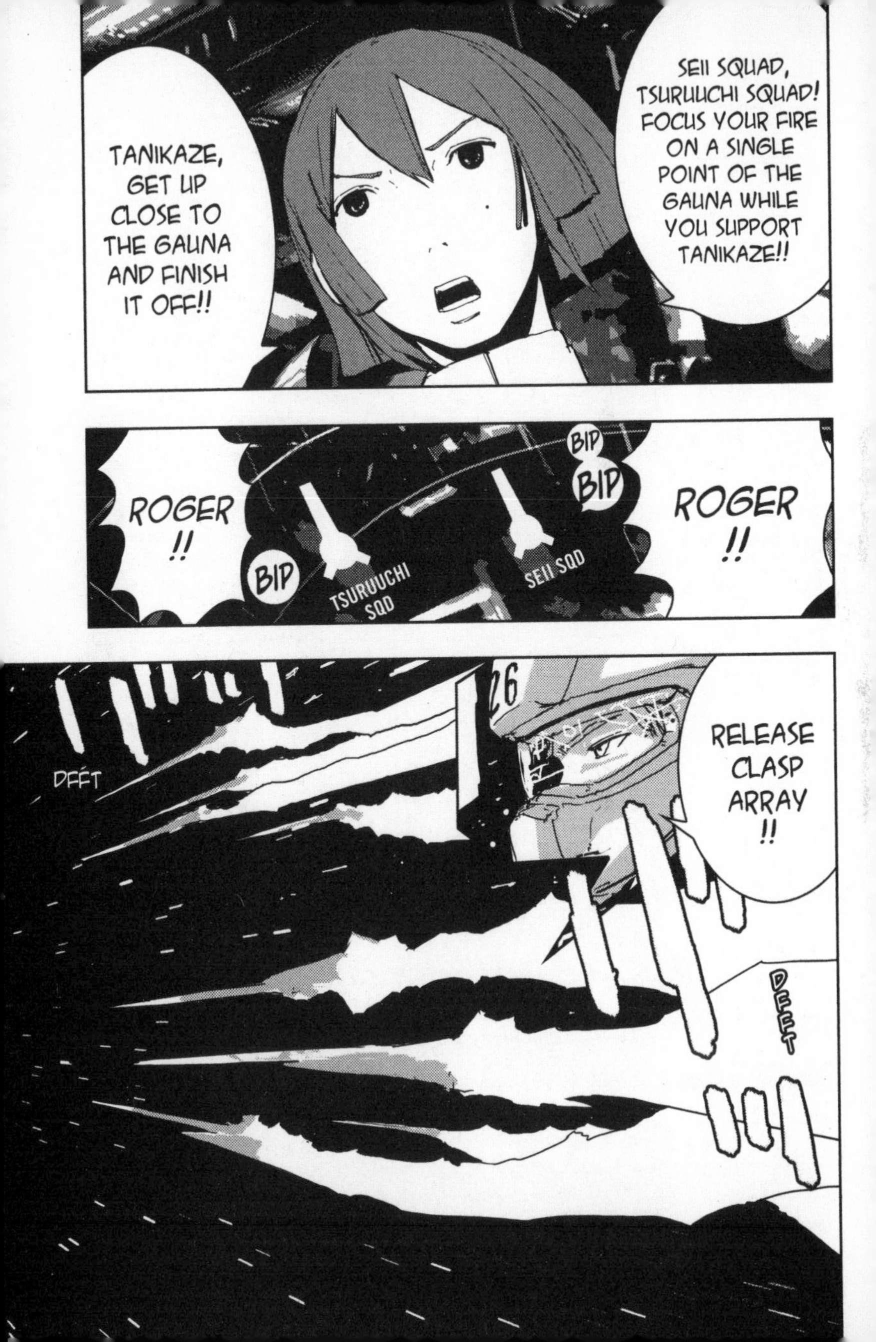
SEII SQUAD, TSURUUCHI SQUAD! FOCUS YOUR FIRE ON A SINGLE POINT OF THE GAUNA WHILE YOU SUPPORT TANIKAZE!!
TANIKAZE, GET UP CLOSE TO THE GAUNA AND FINISH IT OFF!!
ROGER!!
BIP
BIP
SEII SQD
TSURUUCHI SQD
BIP
ROGER!!
26
RELEASE CLASP ARRAY!!
DFFT
DFFT

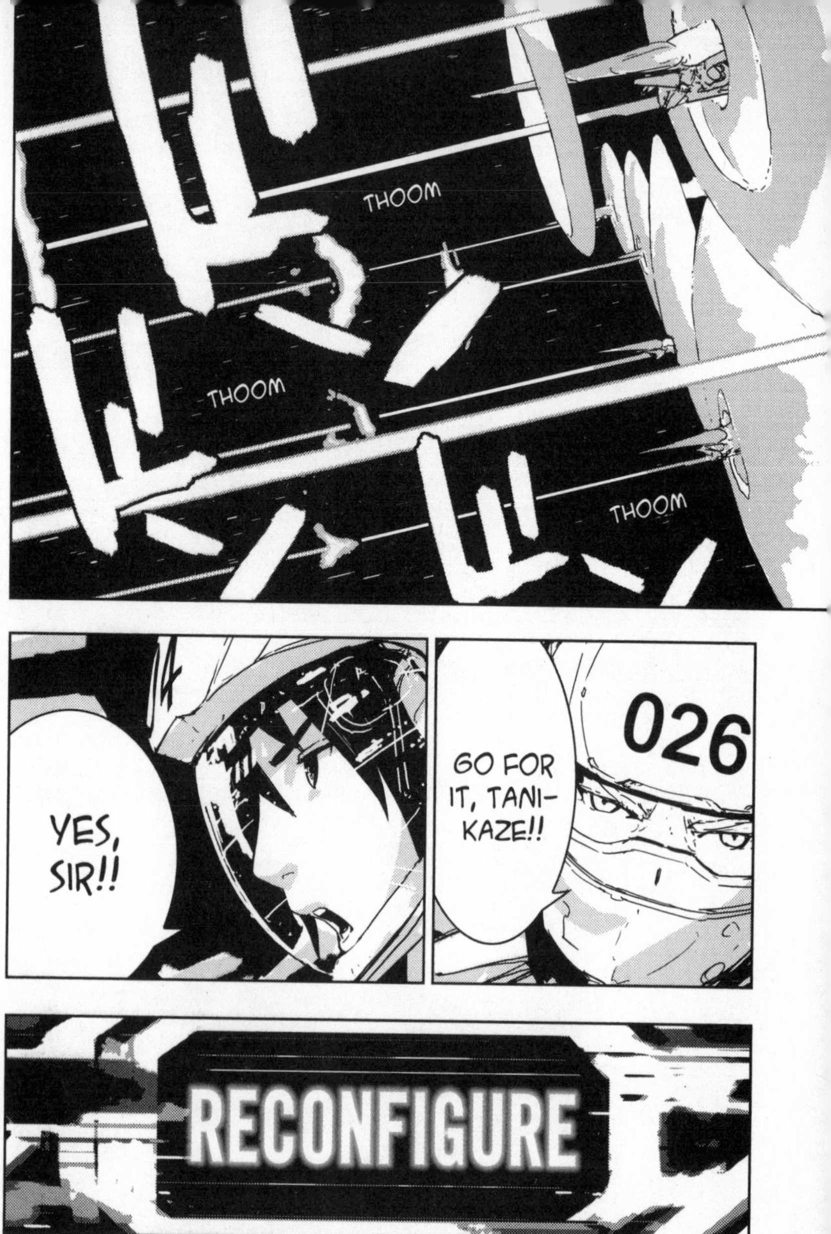
THOOM
THOOM
THOOM
026
GO FOR IT, TANI-KAZE!!
YES, SIR!!
RECONFIGURE

GWUSH
CHKING
CHKING

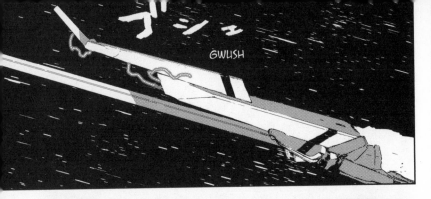

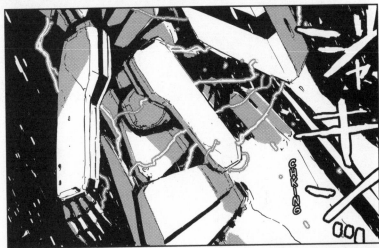

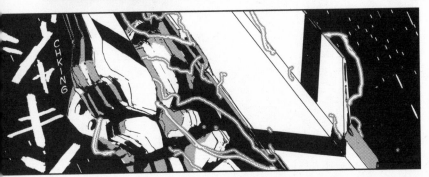

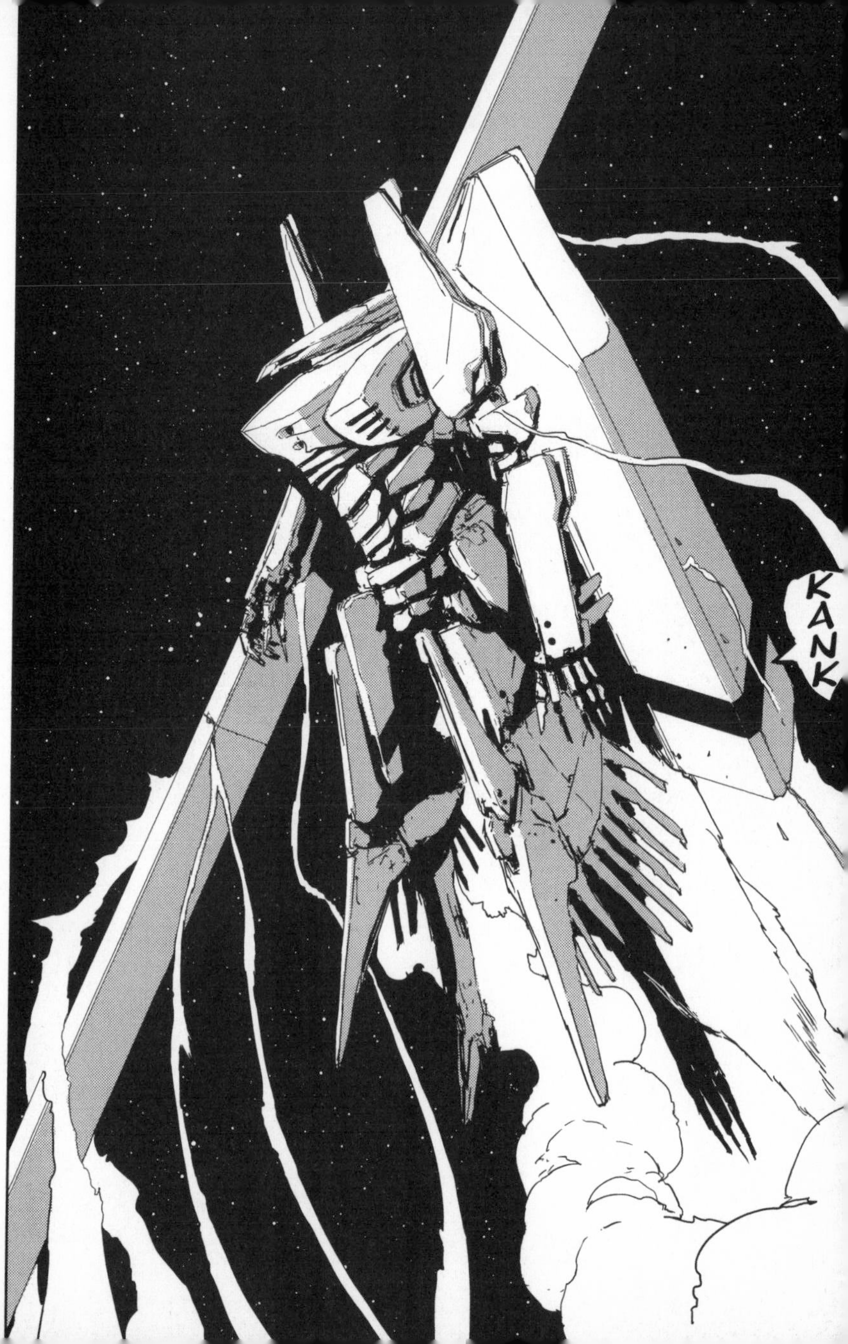
KANK

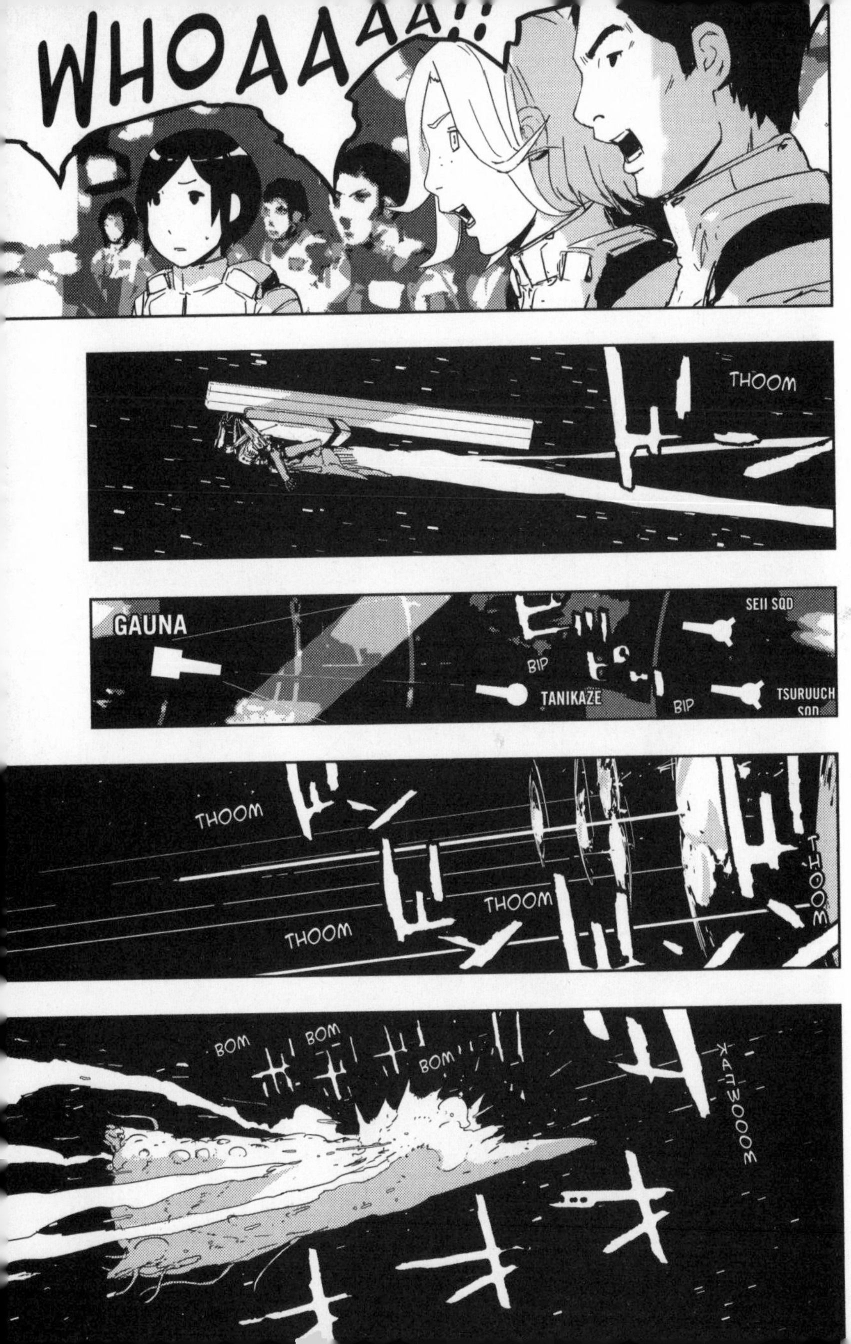
WHOAAAA!!
THOOM
SEII SQD
TSURUUCH
SQD
BIP
TANIKAZE
BIP
GAUNA
THOOM
THOOM
THOOM
THOOM
KATWOOOM
BOM
BOM
BOM

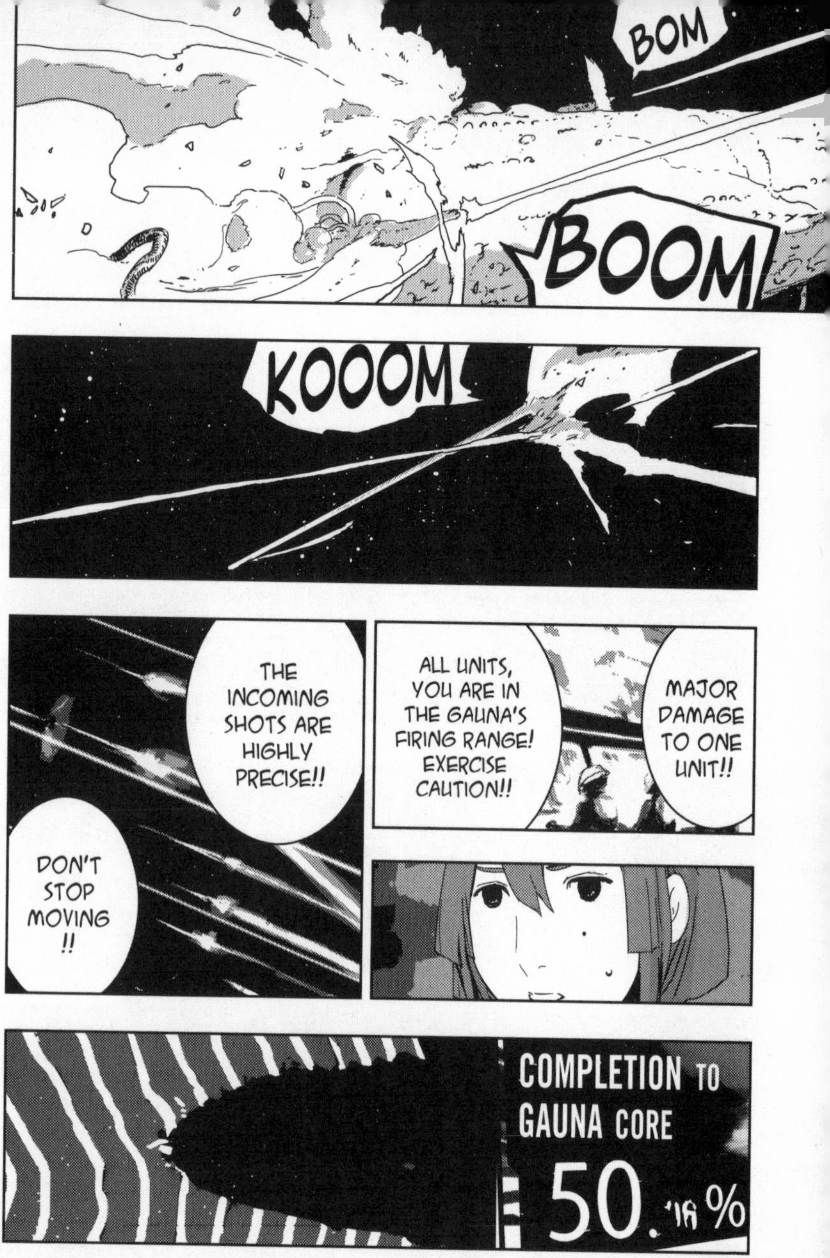
BOM
BOOM
KOOOM
MAJOR DAMAGE TO ONE UNIT!!
ALL UNITS, YOU ARE IN THE GAUNA'S FIRING RANGE! EXERCISE CAUTION!!
THE INCOMING SHOTS ARE HIGHLY PRECISE!!
DON'T STOP MOVING !!
COMPLETION TO GAUNA CORE
50. %

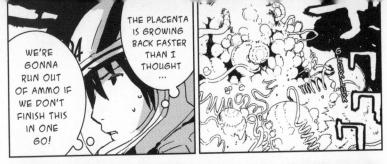

GLUKLUK
THE PLACENTA IS GROWING BACK FASTER THAN I THOUGHT ...
WE'RE GONNA RUN OUT OF AMMO IF WE DON'T FINISH THIS IN ONE GO!
KRACH
HAAAH
HAAH
NAGATE'S NOT DOING WELL...
WHA ?!
HE'S NOT HIS USUAL SELF!
THAT UNIT'S NOT RIGHT FOR HIM ...

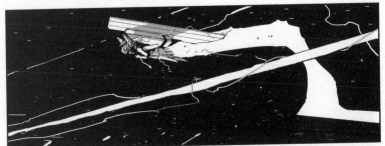

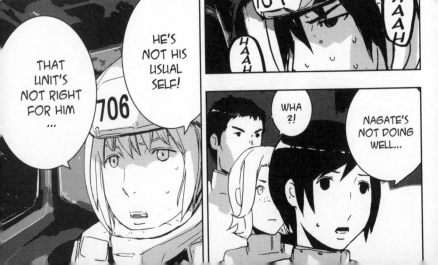

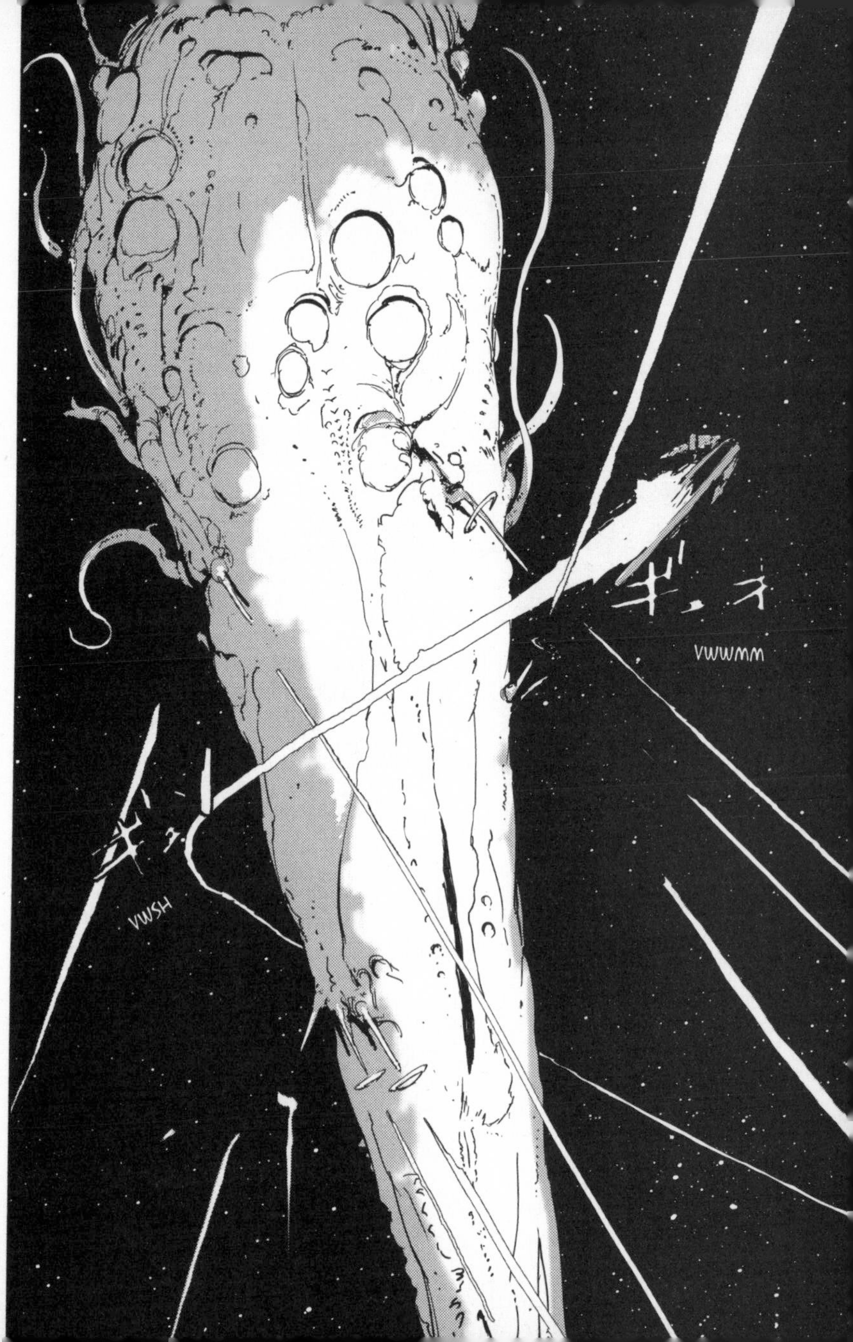
VWUMM
VWSH

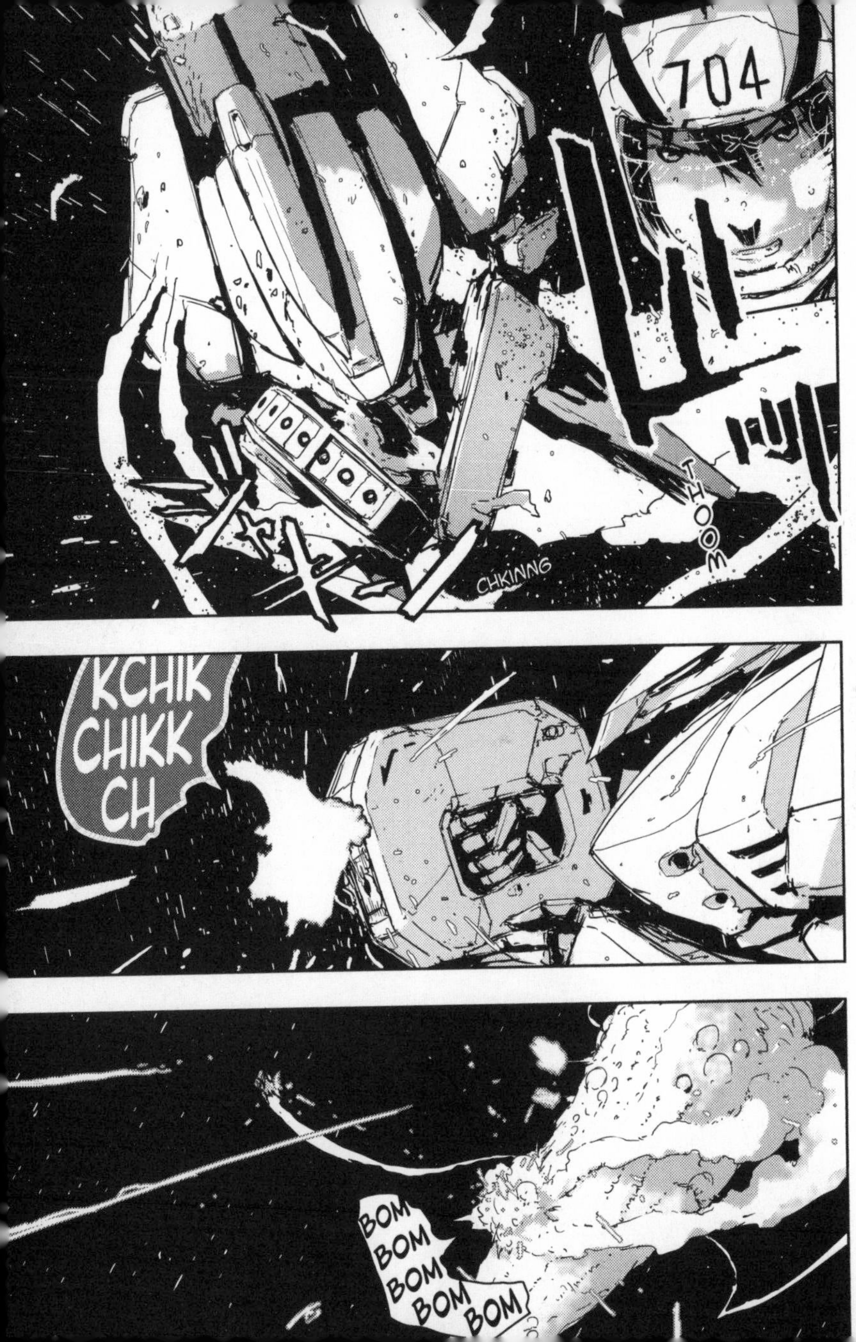
704
THOOOM
CHKINNG
KCHIK
CHIKK
CH
BOM
BOM
BOM
BOM
BOM

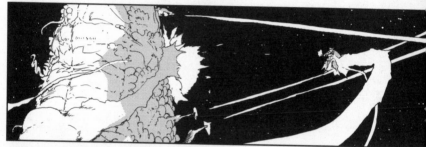

TANIKAZE UNIT
AGD36

AMMO

5/36

COMPLETION TO
GAUNA CORE

80.00%

704

BOOM

BANG

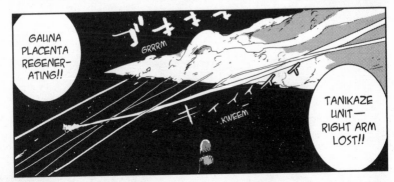

GALUNA PLACENTA REGENER- ATING!!

GRRRM

KWEEM

TANIKAZE UNIT— RIGHT ARM LOST!!

WHAT ABOUT THAT NEXT-GEN HIGGS PARTICLE CANNON ?

NOW! FIRE THE HIGGS PARTICLE CANNON !!

COULDN'T EQUIP IT IN TIME?! ACK, THE UNIT'S GOT NO HEAD CANNON!!

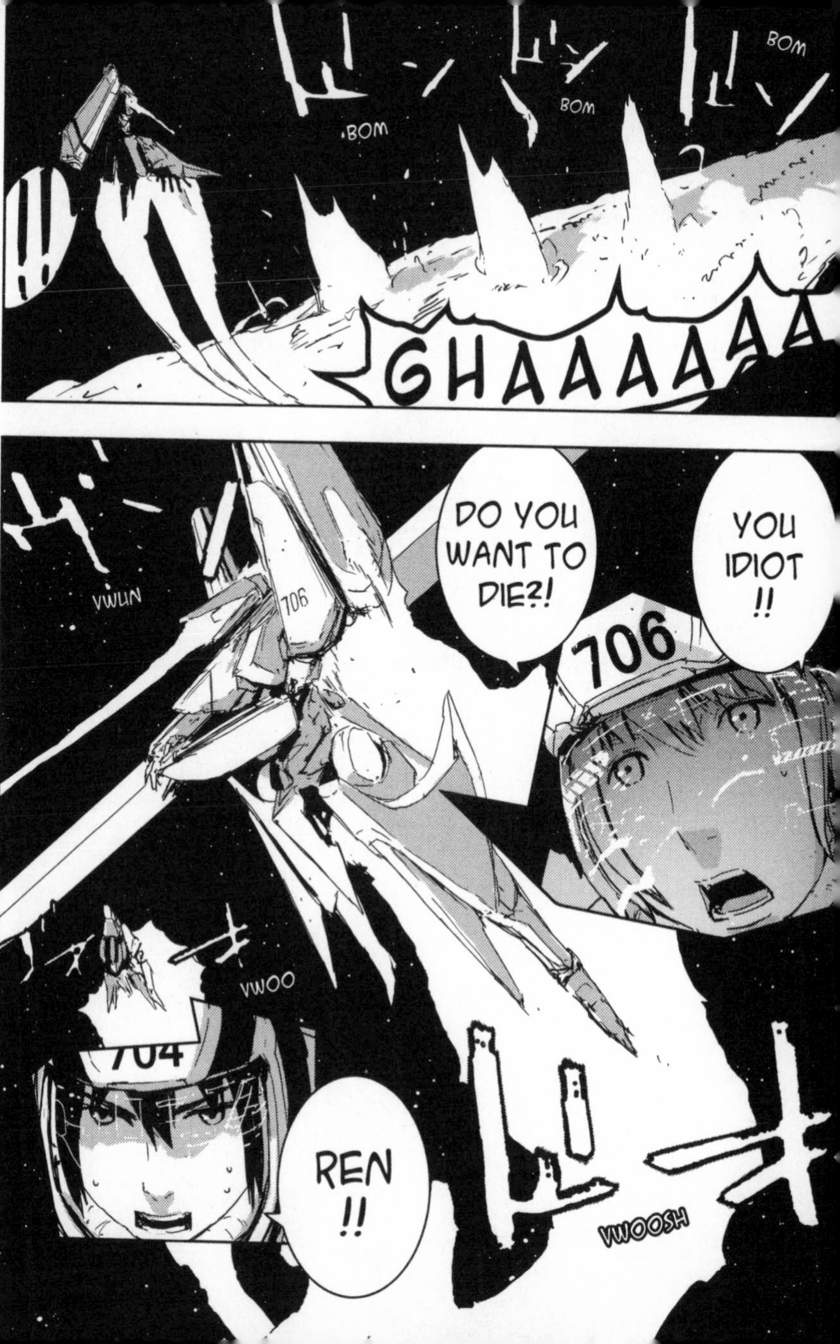
BOM
BOM
BOM
!!
GHAAAAAA
YOU IDIOT !!
DO YOU WANT TO DIE?!
706
VWUN
706
VWOO
704
VWOOSH
REN !!

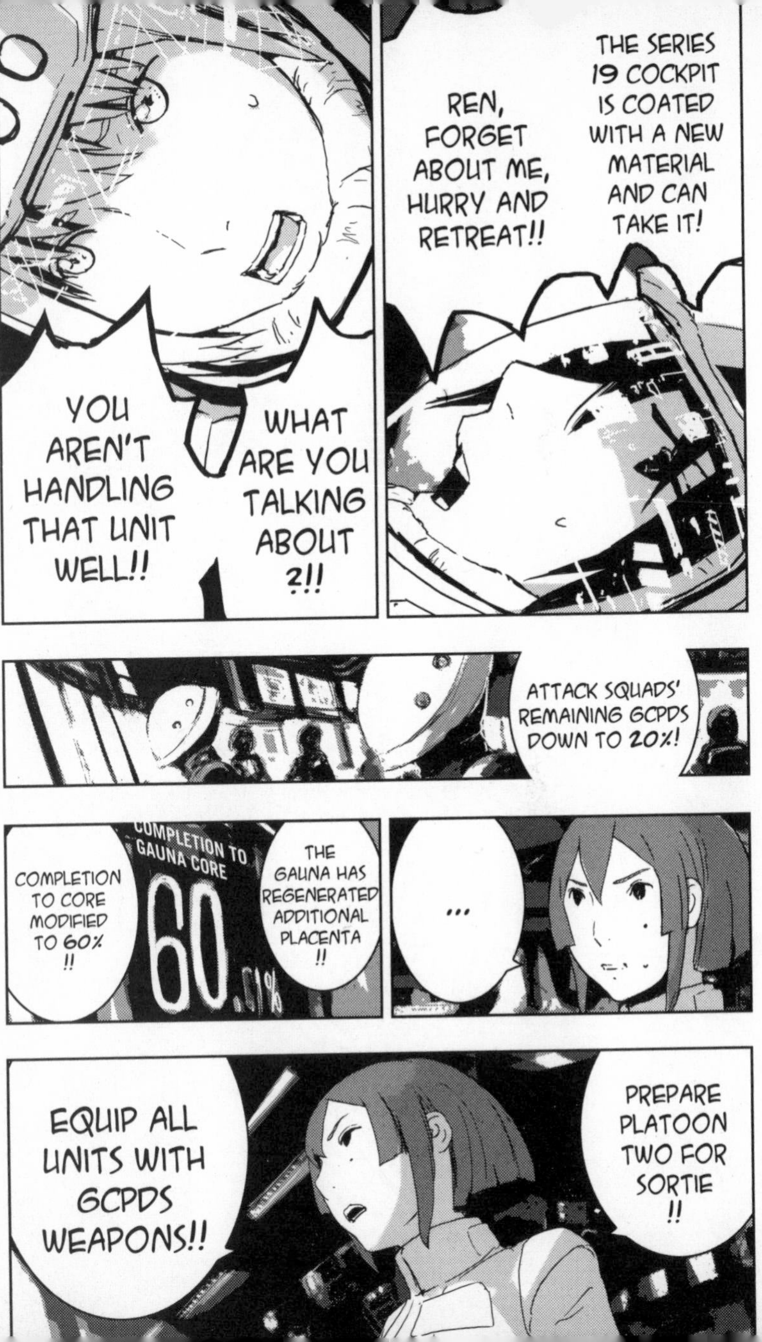
THE SERIES 19 COCKPIT IS COATED WITH A NEW MATERIAL AND CAN TAKE IT!
REN, FORGET ABOUT ME, HURRY AND RETREAT!!
WHAT ARE YOU TALKING ABOUT ?!!
YOU AREN'T HANDLING THAT UNIT WELL!!
ATTACK SQUADS' REMAINING GCPDS DOWN TO 20%!
...
THE GAUNA HAS REGENERATED ADDITIONAL PLACENTA !!
COMPLETION TO GAUNA CORE
60.61%
COMPLETION TO CORE MODIFIED TO 60% !!
PREPARE PLATOON TWO FOR SORTIE !!
EQUIP ALL UNITS WITH GCPDS WEAPONS!!

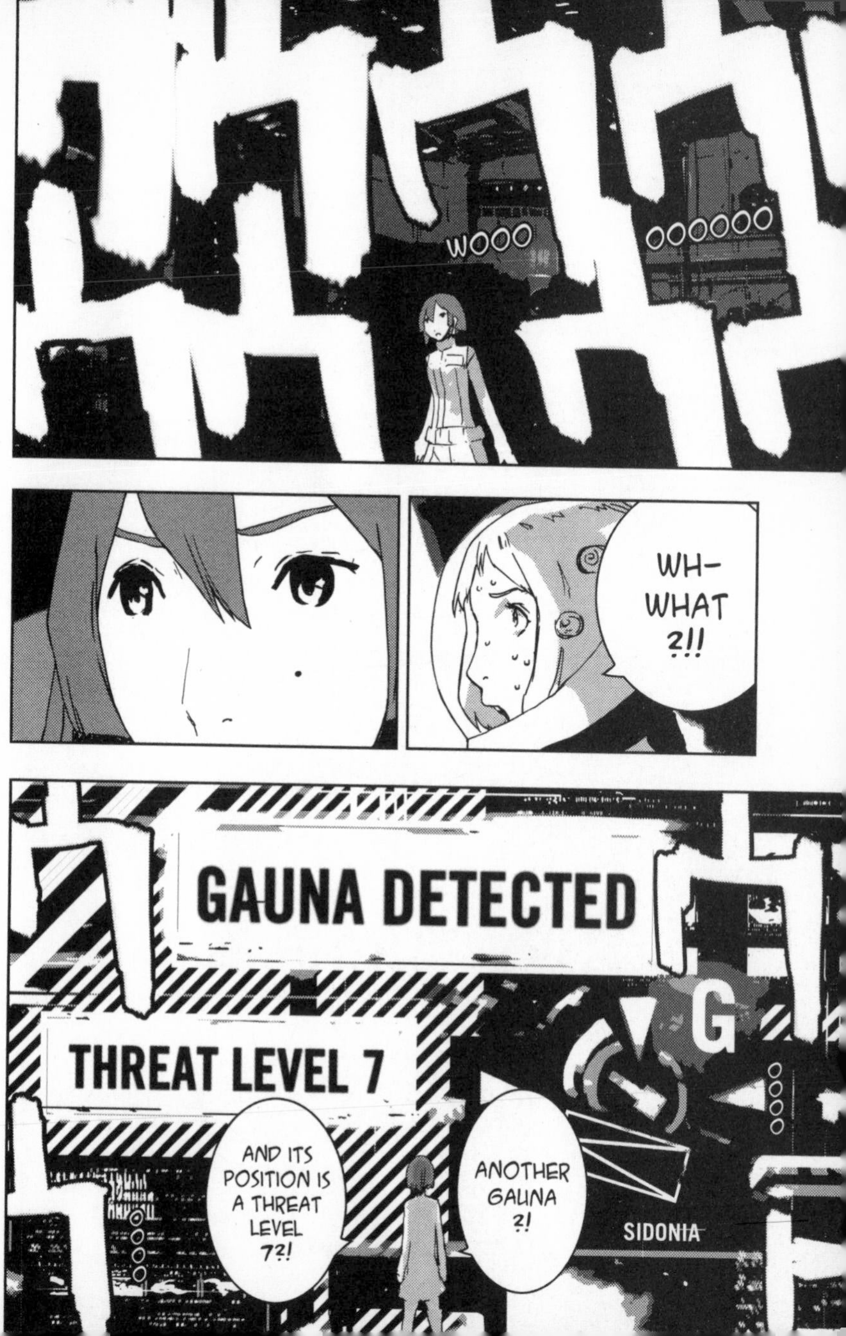
WOOO OOOOOO
WH-WHAT?!!
GAUNA DETECTED
G
THREAT LEVEL 7
SIDONIA
ANOTHER GAUNA?!
AND ITS POSITION IS A THREAT LEVEL 7?!

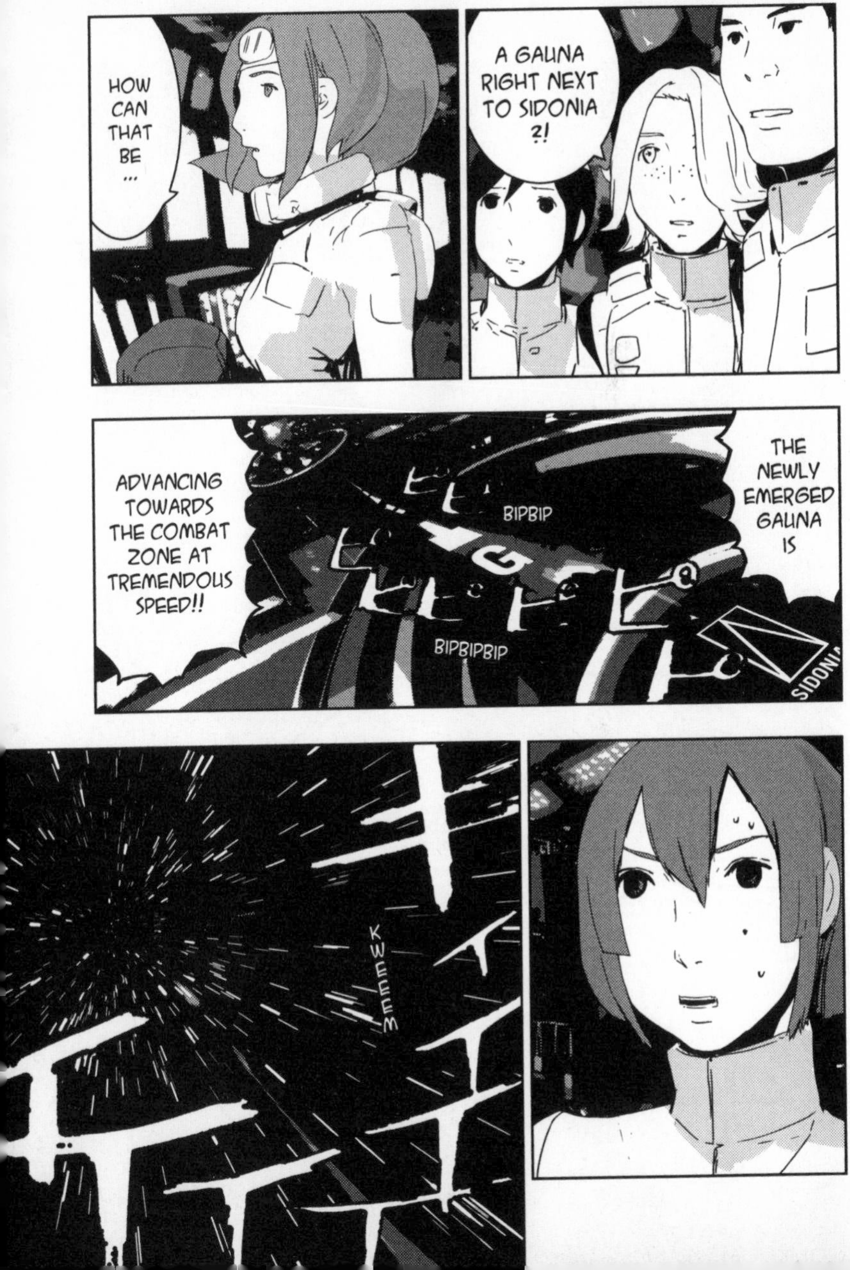
A GAUNA RIGHT NEXT TO SIDONIA ?!
HOW CAN THAT BE ...
THE NEWLY EMERGED GAUNA IS
BIPBIP
ADVANCING TOWARDS THE COMBAT ZONE AT TREMENDOUS SPEED!!
BIPBIPBIP
SIDONIA
KWEEEEM

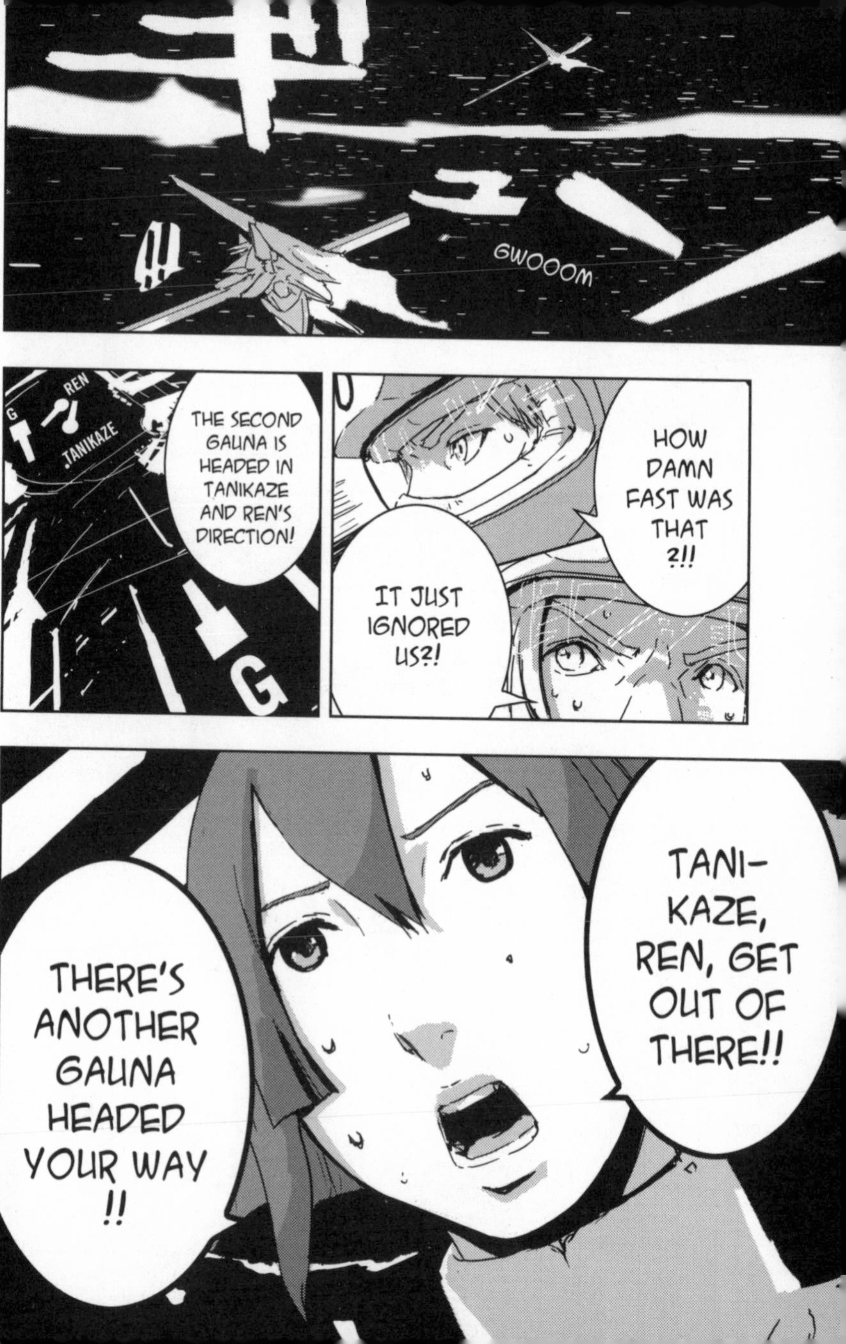
GWOOOM
HOW DAMN FAST WAS THAT ?!!
IT JUST IGNORED US?!
THE SECOND GAUNA IS HEADED IN TANIKAZE AND REN'S DIRECTION!
REN
G
TANIKAZE
G
TANI-KAZE, REN, GET OUT OF THERE!!
THERE'S ANOTHER GAUNA HEADED YOUR WAY !!

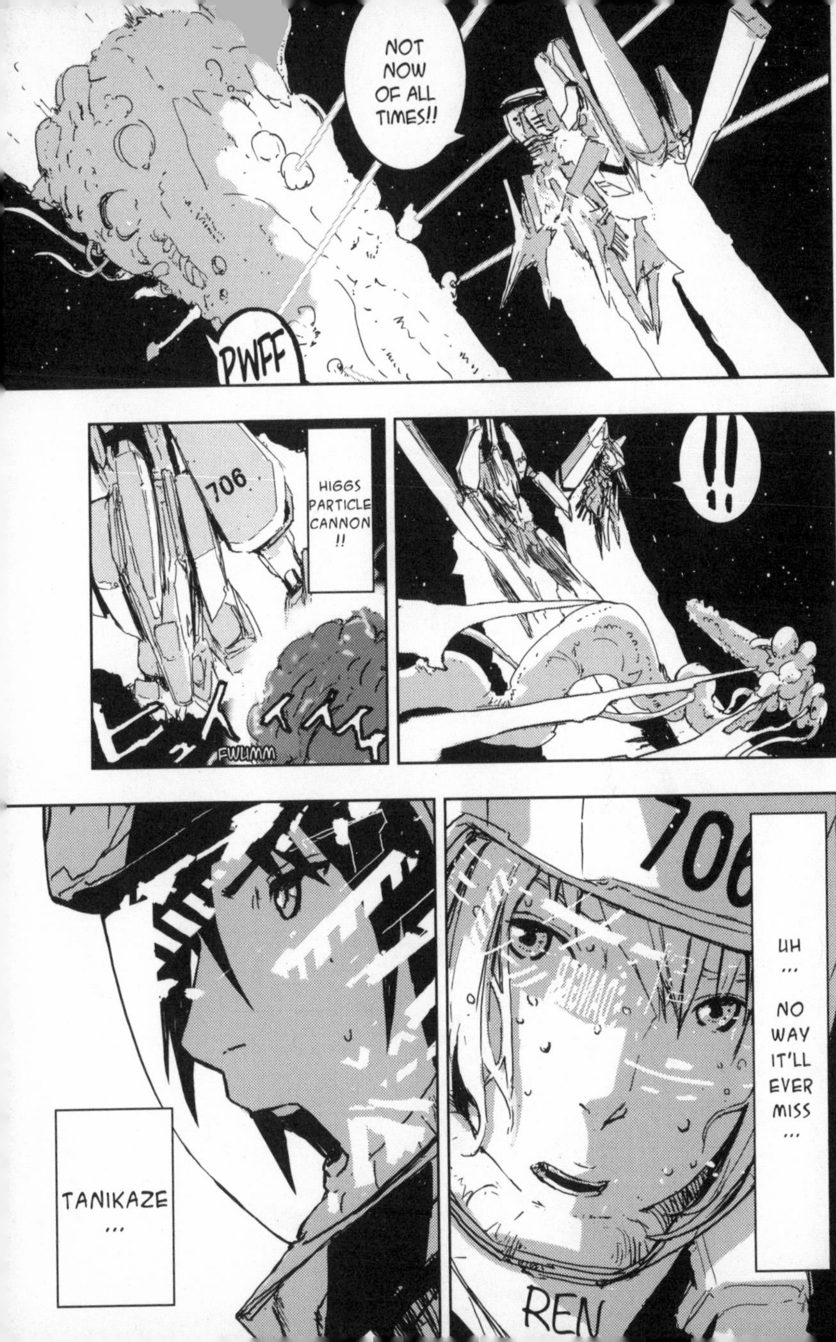
NOT NOW OF ALL TIMES!!
PWFF
!!
HIGGS PARTICLE CANNON !!
706
FWUMM
706
UH ...
NO WAY IT'LL EVER MISS ...
REN
TANIKAZE ...

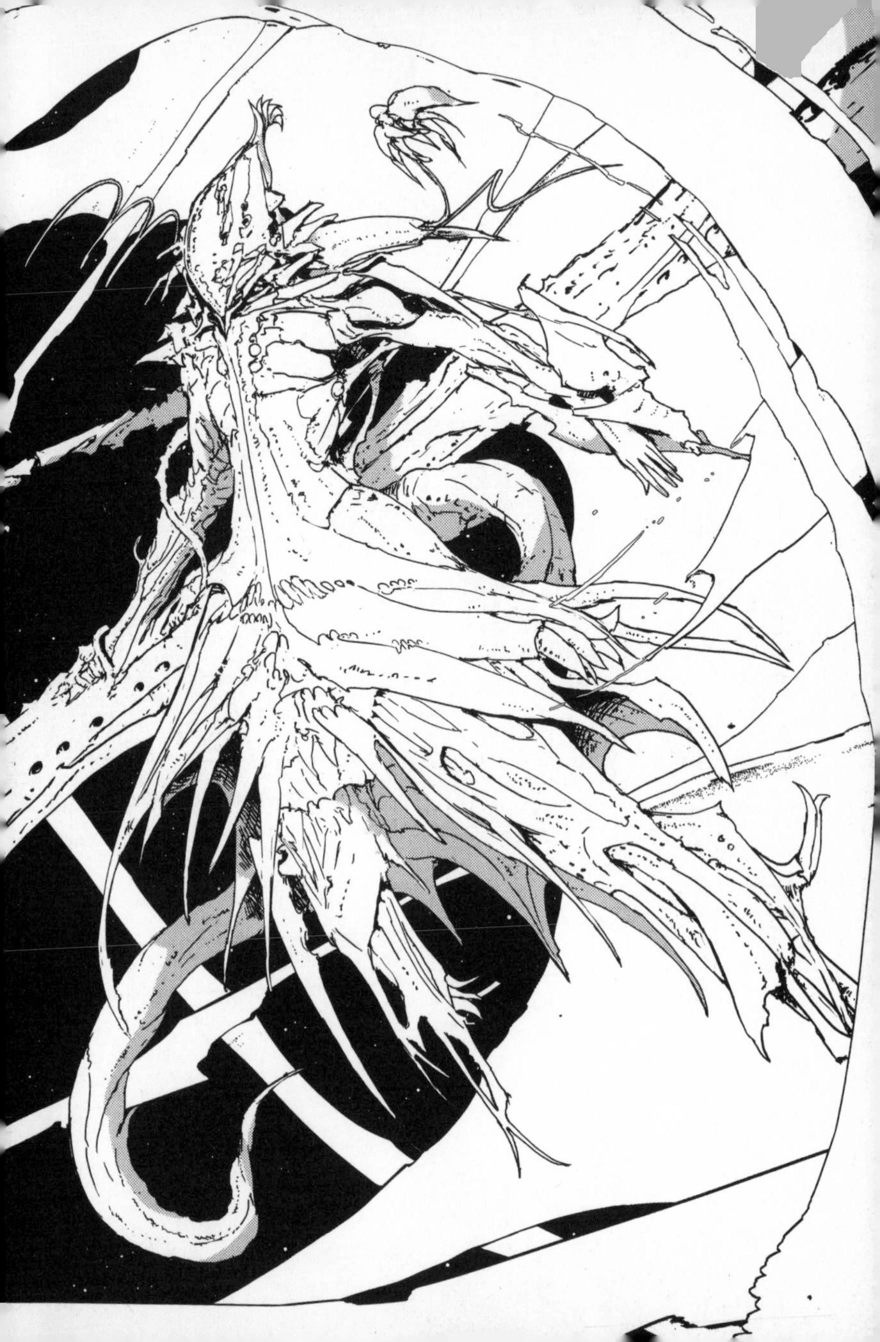

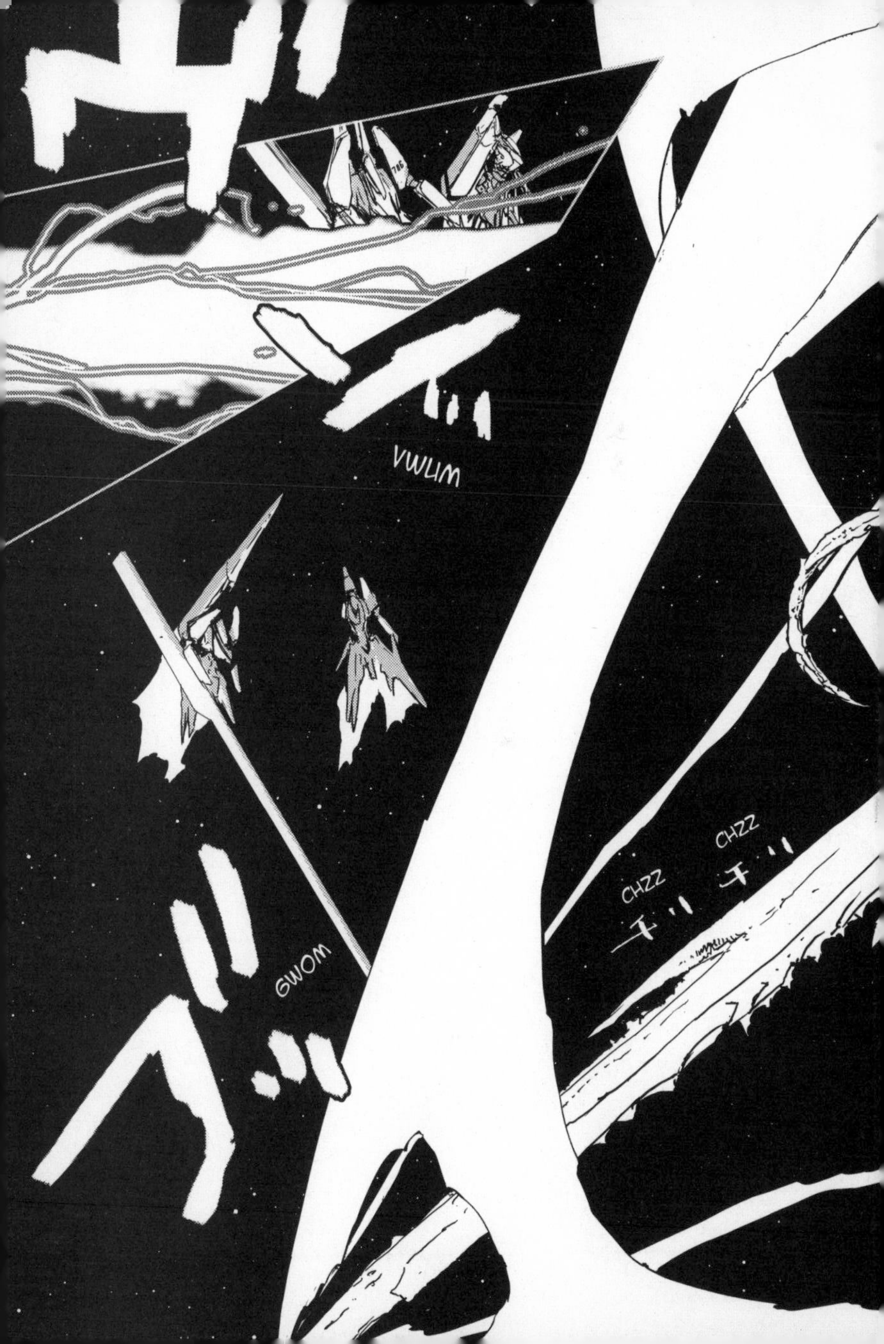
VWUM
CHZZ
CHZZ
GWOM

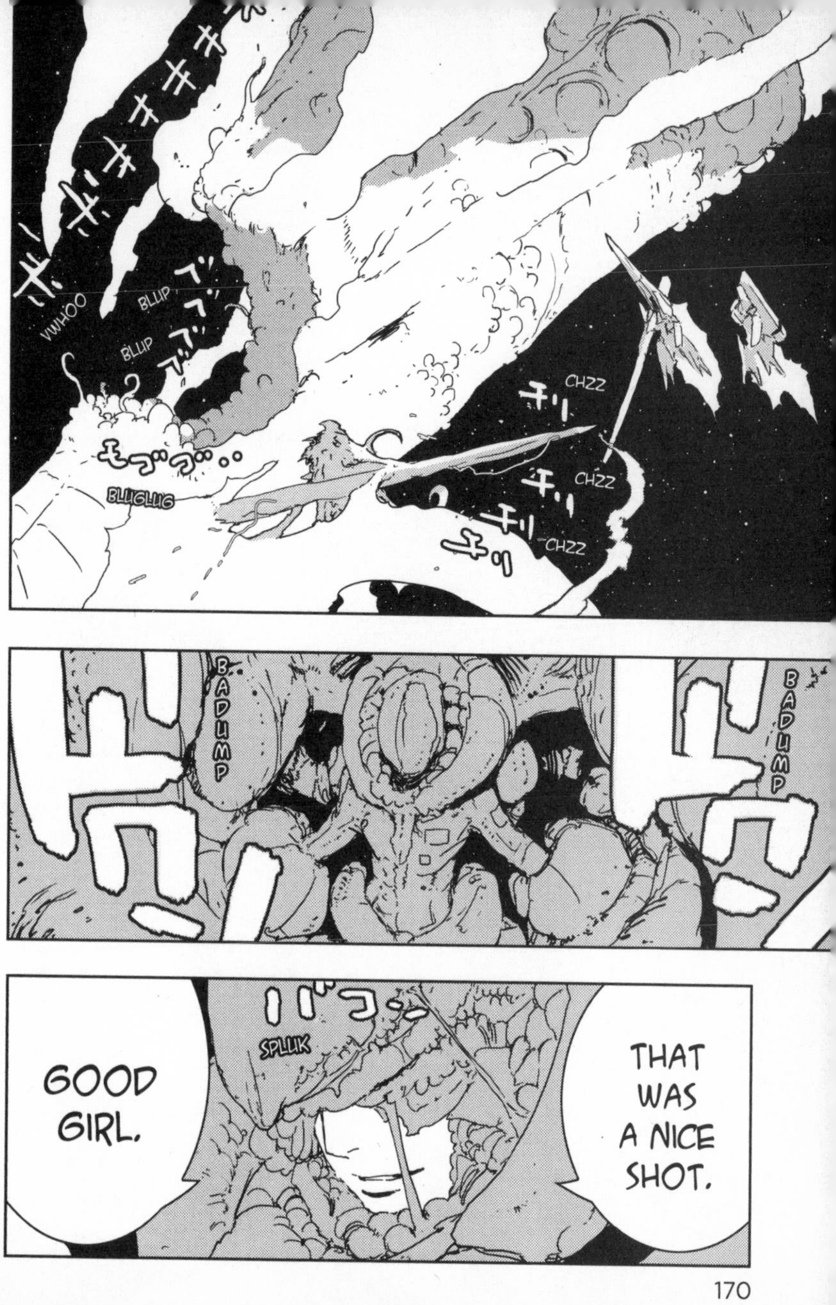
CHZZ
CHZZ
CHZZ
VWHOO
BLLIP
BLLIP
BLUGLUG
BADUMP
BADUMP
SPLUK
THAT WAS A NICE SHOT.
GOOD GIRL.

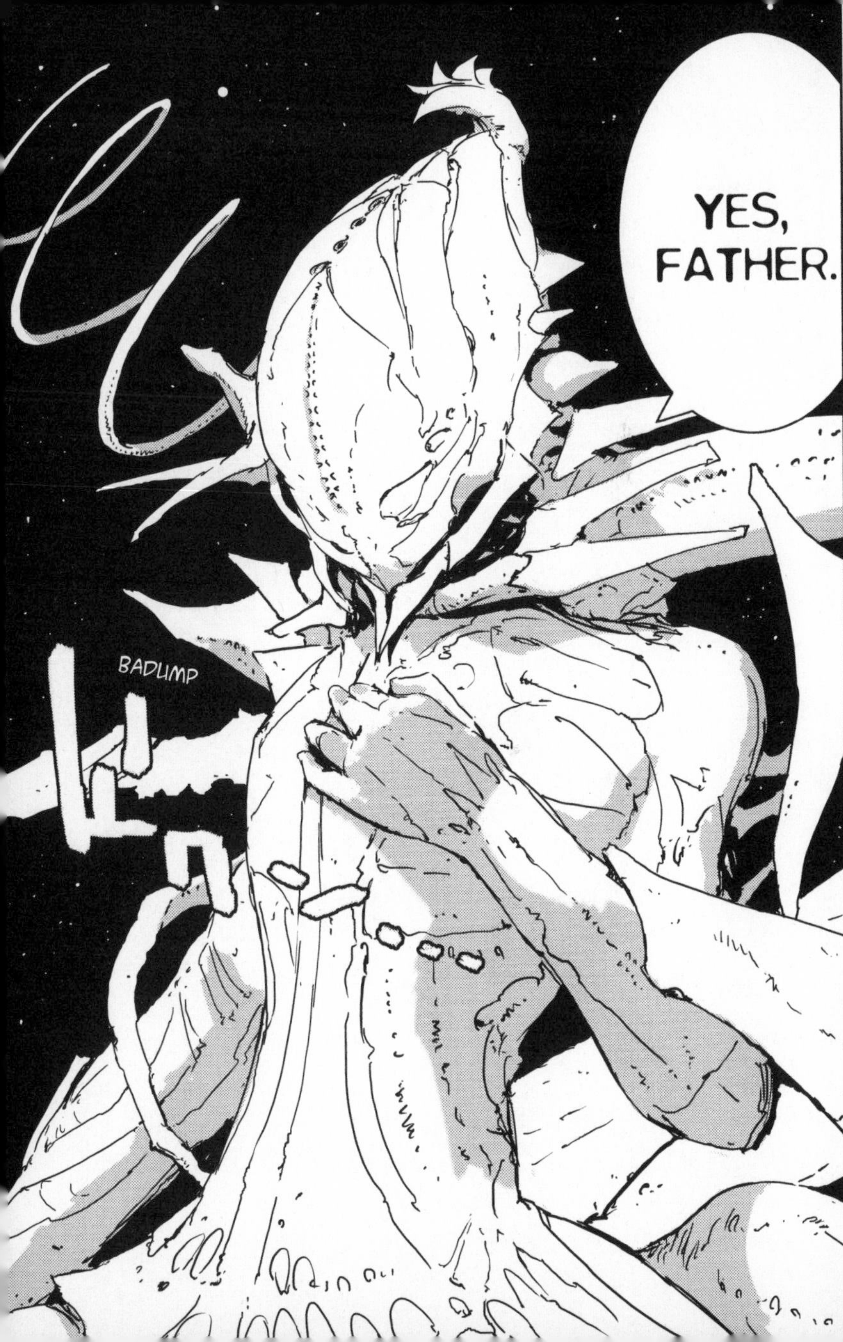
YES, FATHER.
BADUMP
ドク

KNIGHTS OF SIDONIA Volume ⑥: END

附録
SUPPLEMENT
KNIGHTS OF SIDONIA

Nibinben* Goes Forth! Manga Field Journal

Vienna Arc Part 1 By: Tsutomu Nihei

* Nibinben = alternate reading of the characters for "Nihei Tsutomu"

VOTA MEA REDDAM IN CONSPECTV CIMENTIVM OBYA

THE FAMOUS TOURIST ATTRACTION ST. CHARLES'S CHURCH IN VIENNA HAS BEEN TAKEN OVER BY AUSTRIANS COSTUMED AS JAPANESE MANGA AND ANIME CHARACTERS.

THE SIGHTSEERS DON'T HAVE A CLUE WHAT'S GOING ON— HILARIOUS!

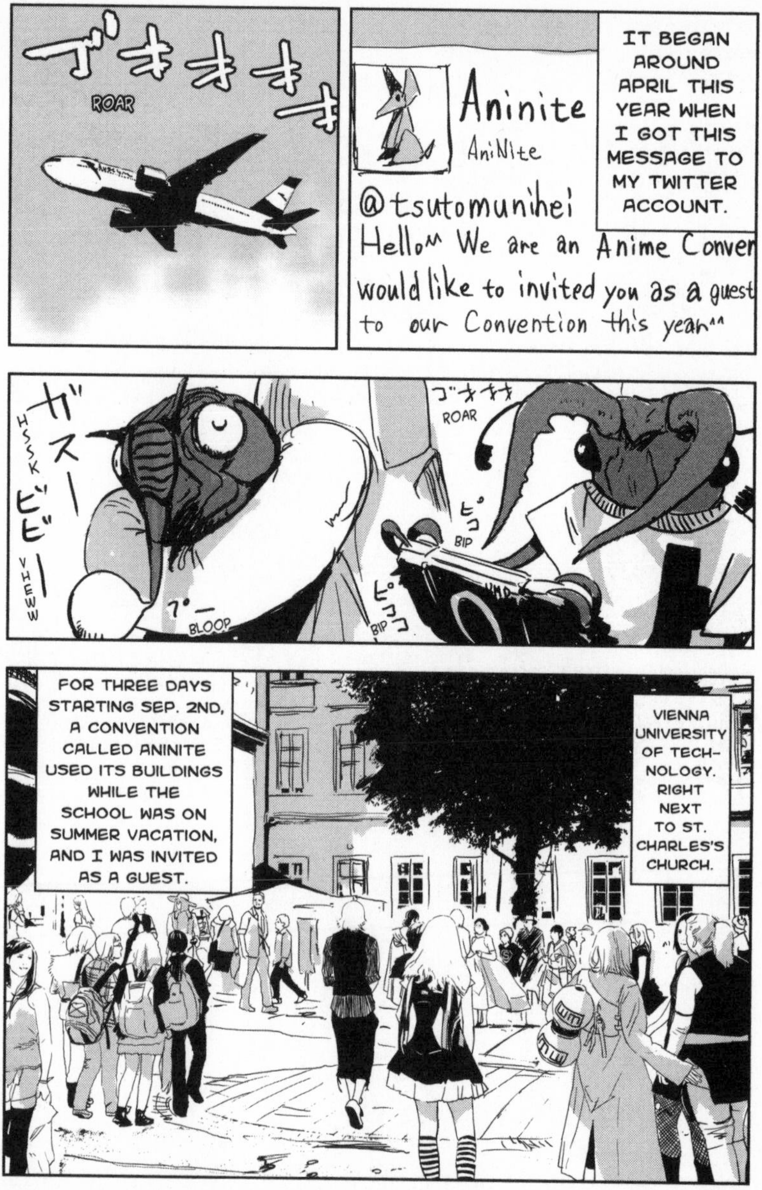
IT BEGAN AROUND APRIL THIS YEAR WHEN I GOT THIS MESSAGE TO MY TWITTER ACCOUNT.
Aninite
AniNite
@tsutomunihei
Hello^^ We are an Anime Conver
would like to invited you as a guest
to our Convention this year^^
ゴオォォォ
ROAR
ゴオォォ
ROAR
ピコ
BIP
ガスー
HSSK
ビビー
VHEWW
ピココ
BIP
プー
BLOOP
VIENNA UNIVERSITY OF TECH-NOLOGY. RIGHT NEXT TO ST. CHARLES'S CHURCH.
FOR THREE DAYS STARTING SEP. 2ND, A CONVENTION CALLED ANINITE USED ITS BUILDINGS WHILE THE SCHOOL WAS ON SUMMER VACATION, AND I WAS INVITED AS A GUEST.

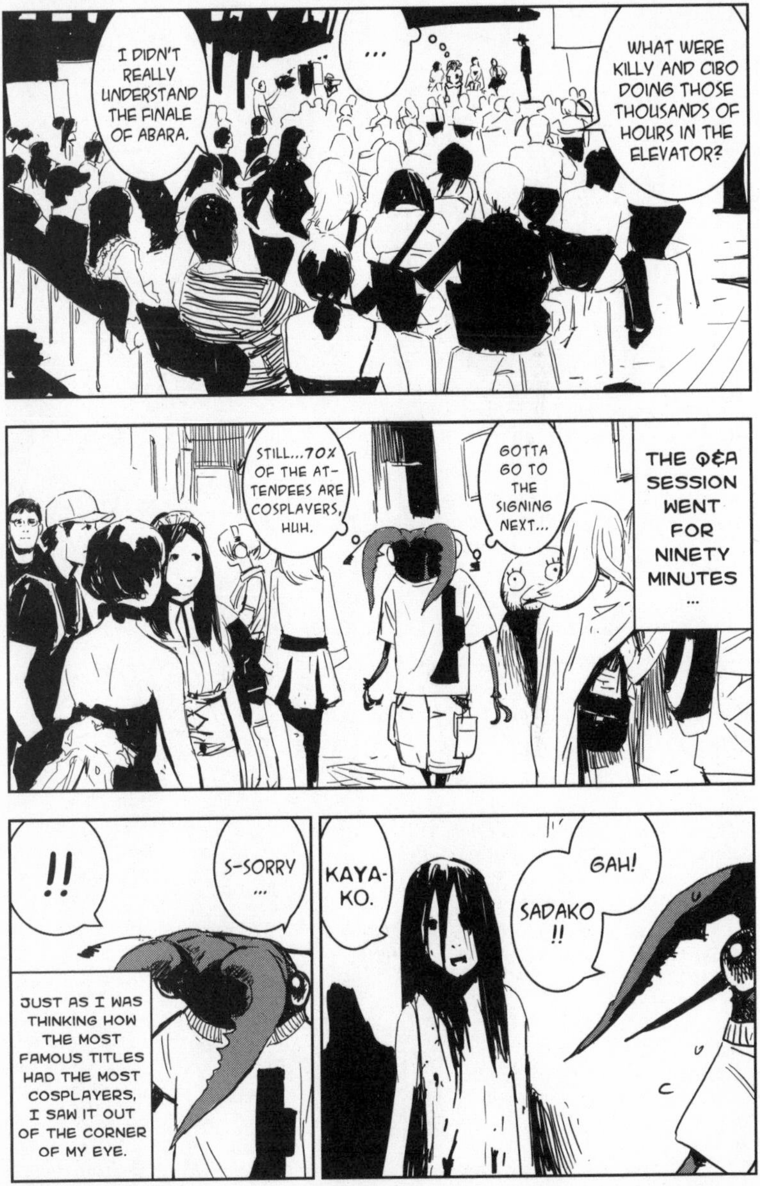
WHAT WERE KILLY AND CIBO DOING THOSE THOUSANDS OF HOURS IN THE ELEVATOR?
...
I DIDN'T REALLY UNDERSTAND THE FINALE OF ABARA.
THE Q&A SESSION WENT FOR NINETY MINUTES ...
GOTTA GO TO THE SIGNING NEXT...
STILL...70% OF THE ATTENDEES ARE COSPLAYERS, HUH.
GAH!
SADAKO !!
KAYA-KO.
S-SORRY ...
!!
JUST AS I WAS THINKING HOW THE MOST FAMOUS TITLES HAD THE MOST COSPLAYERS, I SAW IT OUT OF THE CORNER OF MY EYE.

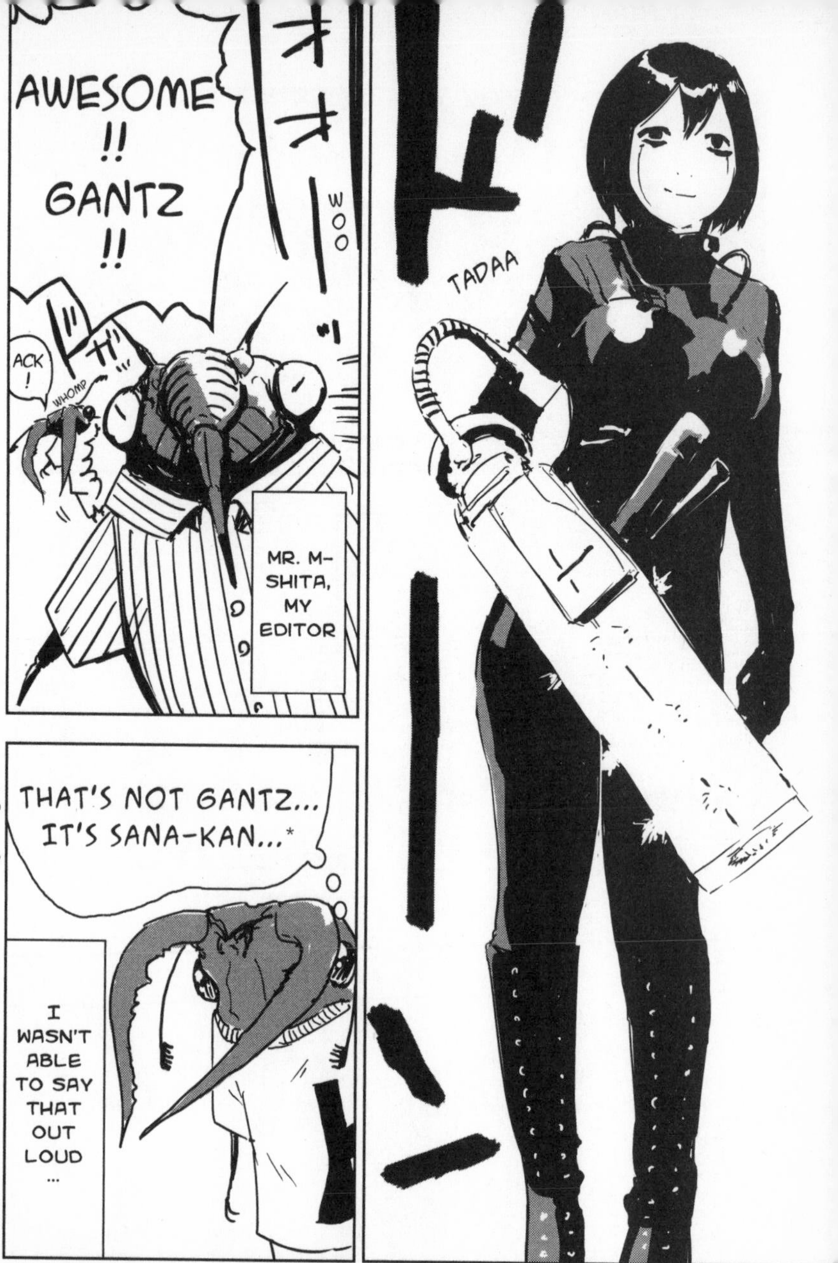
TADAA
AWESOME!! GANTZ!!
WOO
ACK!
WHOMP
MR. M-SHITA, MY EDITOR
THAT'S NOT GANTZ... IT'S SANA-KAN...*
I WASN'T ABLE TO SAY THAT OUT LOUD ...

*A character who appeared in Tsutomu Nihei's debut manga "BLAME!" Her right arm/weapon was a distinguishing feature.

CONTINUED IN PART ②

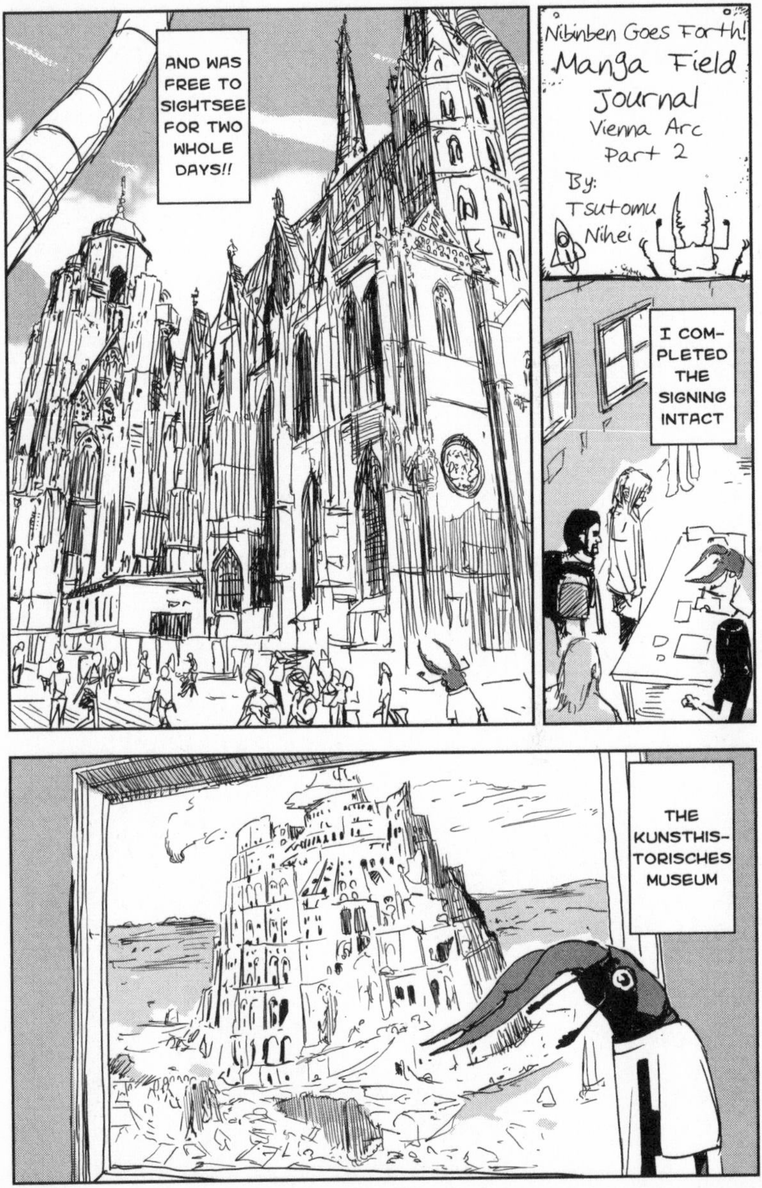
Nibinben Goes Forth!
Manga Field Journal
Vienna Arc Part 2
By: Tsutomu Nihei
I COMPLETED THE SIGNING INTACT
AND WAS FREE TO SIGHTSEE FOR TWO WHOLE DAYS!!
THE KUNSTHISTORISCHES MUSEUM

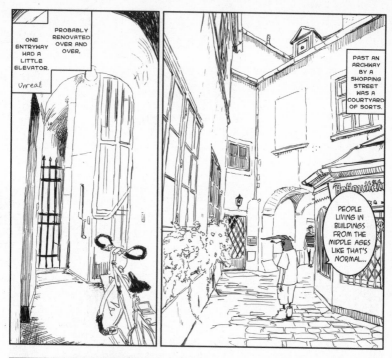

PAST AN ARCHWAY BY A SHOPPING STREET WAS A COURTYARD OF SORTS.
PEOPLE LIVING IN BUILDINGS FROM THE MIDDLE AGES LIKE THAT'S NORMAL...
PROBABLY RENOVATED OVER AND OVER,
ONE ENTRYWAY HAD A LITTLE ELEVATOR.
Unreal!
THANK YOU, INTERPRETERS KLARA AND YURI AND PROMOTER CHANG!!
BUZZZ
BUZZZ
VIENNA WAS THE BEST!!

MANGA FIELD JOURNAL VIENNA ARC Part② END

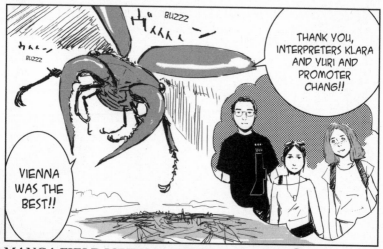

KNIGHTS OF SIDONIA Volume ⑦
ON SALE NOW!!

Knights of Sidonia, volume 6

Translation: Kumar Sivasubramanian
Production: Grace Lu
Daniela Yamada
Anthony Quintessenza

First published in Japan in 2011 by Kodansha, Ltd., Tokyo
Publication for this English edition arranged through Kodansha, Ltd., Tokyo
English language version produced by Vertical, Inc.

Translation provided by Vertical, Inc., 2013
Published by Vertical, Inc., New York

Originally published in Japanese as *Shidonia no Kishi 6* by Kodansha, Ltd.
Shidonia no Kishi first serialized in *Afternoon*, Kodansha, Ltd., 2009-

ISBN: 978-1-932234-91-6

Manufactured in Canada

First Edition

Second Printing

Vertical, Inc.
451 Park Avenue South
7th Floor
New York, NY 10016
www.vertical-inc.com